Reviews of Environmental Contamination and Toxicology

VOLUME 232

For further volumes:
http://www.springer.com/series/398

Reviews of Environmental Contamination and Toxicology

Editor
David M. Whitacre

Editorial Board
Maria Fernanda, Cavieres, Valparaiso, Chile • Charles P. Gerba, Tucson, Arizona, USA
John Giesy, Saskatoon, Saskatchewan, Canada • O. Hutzinger, Bayreuth, Germany
James B. Knaak, Getzville, New York, USA
James T. Stevens, Winston-Salem, North Carolina, USA
Ronald S. Tjeerdema, Davis, California, USA • Pim de Voogt, Amsterdam, The Netherlands
George W. Ware, Tucson, Arizona, USA

Founding Editor
Francis A. Gunther

VOLUME 232

Coordinating Board of Editors

Dr. David M. Whitacre, *Editor*
Reviews of Environmental Contamination and Toxicology

5115 Bunch Road
Summerfield, North Carolina 27358, USA
(336) 634-2131 (PHONE and FAX)
E-mail: dmwhitacre@triad.rr.com

Dr. Erin R. Bennett, *Editor*
Bulletin of Environmental Contamination and Toxicology
Great Lakes Institute for Environmental Research

University of Windsor
Windsor, Ontario, Canada
E-mail: ebennett@uwindsor.ca

Peter S. Ross, *Editor*
Archives of Environmental Contamination and Toxicology

Fisheries and Oceans Canada
Institute of Ocean Sciences Sidney
British Colombia, Canada
E-mail: peter.s.ross@dfo-mpo.gc.ca

ISSN 0179-5953   ISSN 2197-6554 (electronic)
ISBN 978-3-319-06745-2   ISBN 978-3-319-06746-9 (eBook)
DOI 10.1007/978-3-319-06746-9
Springer Cham Heidelberg New York Dordrecht London

© Springer International Publishing Switzerland 2014
This work is subject to copyright. All rights are reserved by the Publisher, whether the whole or part of the material is concerned, specifically the rights of translation, reprinting, reuse of illustrations, recitation, broadcasting, reproduction on microfilms or in any other physical way, and transmission or information storage and retrieval, electronic adaptation, computer software, or by similar or dissimilar methodology now known or hereafter developed. Exempted from this legal reservation are brief excerpts in connection with reviews or scholarly analysis or material supplied specifically for the purpose of being entered and executed on a computer system, for exclusive use by the purchaser of the work. Duplication of this publication or parts thereof is permitted only under the provisions of the Copyright Law of the Publisher's location, in its current version, and permission for use must always be obtained from Springer. Permissions for use may be obtained through RightsLink at the Copyright Clearance Center. Violations are liable to prosecution under the respective Copyright Law.
The use of general descriptive names, registered names, trademarks, service marks, etc. in this publication does not imply, even in the absence of a specific statement, that such names are exempt from the relevant protective laws and regulations and therefore free for general use.
While the advice and information in this book are believed to be true and accurate at the date of publication, neither the authors nor the editors nor the publisher can accept any legal responsibility for any errors or omissions that may be made. The publisher makes no warranty, express or implied, with respect to the material contained herein.

Printed on acid-free paper

Springer is part of Springer Science+Business Media (www.springer.com)

Foreword

International concern in scientific, industrial, and governmental communities over traces of xenobiotics in foods and in both abiotic and biotic environments has justified the present triumvirate of specialized publications in this field: comprehensive reviews, rapidly published research papers and progress reports, and archival documentations. These three international publications are integrated and scheduled to provide the coherency essential for nonduplicative and current progress in a field as dynamic and complex as environmental contamination and toxicology. This series is reserved exclusively for the diversified literature on "toxic" chemicals in our food, our feeds, our homes, recreational and working surroundings, our domestic animals, our wildlife, and ourselves. Tremendous efforts worldwide have been mobilized to evaluate the nature, presence, magnitude, fate, and toxicology of the chemicals loosed upon the Earth. Among the sequelae of this broad new emphasis is an undeniable need for an articulated set of authoritative publications, where one can find the latest important world literature produced by these emerging areas of science together with documentation of pertinent ancillary legislation.

Research directors and legislative or administrative advisers do not have the time to scan the escalating number of technical publications that may contain articles important to current responsibility. Rather, these individuals need the background provided by detailed reviews and the assurance that the latest information is made available to them, all with minimal literature searching. Similarly, the scientist assigned or attracted to a new problem is required to glean all literature pertinent to the task, to publish new developments or important new experimental details quickly, to inform others of findings that might alter their own efforts, and eventually to publish all his/her supporting data and conclusions for archival purposes.

In the fields of environmental contamination and toxicology, the sum of these concerns and responsibilities is decisively addressed by the uniform, encompassing, and timely publication format of the Springer triumvirate:

Reviews of Environmental Contamination and Toxicology [Vol. 1 through 97 (1962–1986) as Residue Reviews] for detailed review articles concerned with any aspects of chemical contaminants, including pesticides, in the total environment with toxicological considerations and consequences.

Bulletin of Environmental Contamination and Toxicology (Vol. 1 in 1966) for rapid publication of short reports of significant advances and discoveries in the fields of air, soil, water, and food contamination and pollution as well as methodology and other disciplines concerned with the introduction, presence, and effects of toxicants in the total environment.

Archives of Environmental Contamination and Toxicology (Vol. 1 in 1973) for important complete articles emphasizing and describing original experimental or theoretical research work pertaining to the scientific aspects of chemical contaminants in the environment.

Manuscripts for Reviews and the Archives are in identical formats and are peer reviewed by scientists in the field for adequacy and value; manuscripts for the Bulletin are also reviewed, but are published by photo-offset from camera-ready copy to provide the latest results with minimum delay. The individual editors of these three publications comprise the joint Coordinating Board of Editors with referral within the board of manuscripts submitted to one publication but deemed by major emphasis or length more suitable for one of the others.

Coordinating Board of Editors

Preface

The role of *Reviews* is to publish detailed scientific review articles on all aspects of environmental contamination and associated toxicological consequences. Such articles facilitate the often complex task of accessing and interpreting cogent scientific data within the confines of one or more closely related research fields.

In the nearly 50 years since *Reviews of Environmental Contamination and Toxicology* (formerly *Residue Reviews*) was first published, the number, scope, and complexity of environmental pollution incidents have grown unabated. During this entire period, the emphasis has been on publishing articles that address the presence and toxicity of environmental contaminants. New research is published each year on a myriad of environmental pollution issues facing people worldwide. This fact, and the routine discovery and reporting of new environmental contamination cases, creates an increasingly important function for *Reviews*.

The staggering volume of scientific literature demands remedy by which data can be synthesized and made available to readers in an abridged form. *Reviews* addresses this need and provides detailed reviews worldwide to key scientists and science or policy administrators, whether employed by government, universities, or the private sector.

There is a panoply of environmental issues and concerns on which many scientists have focused their research in past years. The scope of this list is quite broad, encompassing environmental events globally that affect marine and terrestrial ecosystems; biotic and abiotic environments; impacts on plants, humans, and wildlife; and pollutants, both chemical and radioactive; as well as the ravages of environmental disease in virtually all environmental media (soil, water, air). New or enhanced safety and environmental concerns have emerged in the last decade to be added to incidents covered by the media, studied by scientists, and addressed by governmental and private institutions. Among these are events so striking that they are creating a paradigm shift. Two in particular are at the center of everincreasing media as well as scientific attention: bioterrorism and global warming. Unfortunately, these very worrisome issues are now superimposed on the already extensive list of ongoing environmental challenges.

The ultimate role of publishing scientific research is to enhance understanding of the environment in ways that allow the public to be better informed. The term "informed public" as used by Thomas Jefferson in the age of enlightenment conveyed the thought of soundness and good judgment. In the modern sense, being "well informed" has the narrower meaning of having access to sufficient information. Because the public still gets most of its information on science and technology from TV news and reports, the role for scientists as interpreters and brokers of scientific information to the public will grow rather than diminish. Environmentalism is the newest global political force, resulting in the emergence of multinational consortia to control pollution and the evolution of the environmental ethic. Will the new politics of the twenty-first century involve a consortium of technologists and environmentalists, or a progressive confrontation? These matters are of genuine concern to governmental agencies and legislative bodies around the world.

For those who make the decisions about how our planet is managed, there is an ongoing need for continual surveillance and intelligent controls to avoid endangering the environment, public health, and wildlife. Ensuring safety-in-use of the many chemicals involved in our highly industrialized culture is a dynamic challenge, for the old, established materials are continually being displaced by newly developed molecules more acceptable to federal and state regulatory agencies, public health officials, and environmentalists.

Reviews publishes synoptic articles designed to treat the presence, fate, and, if possible, the safety of xenobiotics in any segment of the environment. These reviews can be either general or specific, but properly lie in the domains of analytical chemistry and its methodology, biochemistry, human and animal medicine, legislation, pharmacology, physiology, toxicology, and regulation. Certain affairs in food technology concerned specifically with pesticide and other food-additive problems may also be appropriate.

Because manuscripts are published in the order in which they are received in final form, it may seem that some important aspects have been neglected at times. However, these apparent omissions are recognized, and pertinent manuscripts are likely in preparation or planned. The field is so very large and the interests in it are so varied that the editor and the editorial board earnestly solicit authors and suggestions of underrepresented topics to make this international book series yet more useful and worthwhile.

Justification for the preparation of any review for this book series is that it deals with some aspect of the many real problems arising from the presence of foreign chemicals in our surroundings. Thus, manuscripts may encompass case studies from any country. Food additives, including pesticides, or their metabolites that may persist into human food and animal feeds are within this scope. Additionally, chemical contamination in any manner of air, water, soil, or plant or animal life is within these objectives and their purview.

Manuscripts are often contributed by invitation. However, nominations for new topics or topics in areas that are rapidly advancing are welcome. Preliminary communication with the editor is recommended before volunteered review manuscripts are submitted.

Summerfield, NC, USA   David M. Whitacre

Contents

**Heavy-Metal-Induced Reactive Oxygen Species:
Phytotoxicity and Physicochemical Changes in Plants**.............................. 1
Muhammad Shahid, Bertrand Pourrut, Camille Dumat,
Muhammad Nadeem, Muhammad Aslam, and Eric Pinelli

Biological Responses of Agricultural Soils to Fly-Ash Amendment 45
Rajeev Pratap Singh, Bhavisha Sharma, Abhijit Sarkar,
Chandan Sengupta, Pooja Singh, and Mahamad Hakimi Ibrahim

Oil Palm Biomass as an Adsorbent for Heavy Metals............................... 61
Mohammadtaghi Vakili, Mohd Rafatullah, Mahamad Hakimi Ibrahim,
Ahmad Zuhairi Abdullah, Babak Salamatinia, and Zahra Gholami

Environmental Fate and Toxicology of Chlorothalonil 89
April R. Van Scoy and Ronald S. Tjeerdema

**The Distribution, Fate, and Effects of Propylene Glycol
Substances in the Environment**... 107
Robert West, Marcy Banton, Jing Hu, and Joanna Klapacz

Index... 139

Heavy-Metal-Induced Reactive Oxygen Species: Phytotoxicity and Physicochemical Changes in Plants

Muhammad Shahid, Bertrand Pourrut, Camille Dumat, Muhammad Nadeem, Muhammad Aslam, and Eric Pinelli

Contents

1 Introduction 2
2 What Are ROS? 4
3 ROS Production in Plant Metabolism 4
   3.1 Natural Production of ROS in Plants 4
   3.2 Heavy-Metal-Induced Production of ROS in Plants 5
4 Roles of ROS in Plant Metabolism 7
5 Toxic Effects of Heavy-Metal-Induced ROS on Macromolecules in Plants 8
   5.1 Lipid Peroxidation 9
   5.2 DNA Damage 11
   5.3 Protein Damage 12
   5.4 Damage to Plant Carbohydrates 13
   5.5 Interference with Signalling 14

M. Shahid
Department of Environmental Sciences, COMSATS Institute of Information Technology, Vehari 61100, Pakistan

INP-ENSAT, Université de Toulouse, Av. de l'Agrobiopôle, 31326 Castanet-Tolosan, France

EcoLab (Laboratoire d'écologie fonctionnelle), INP-ENSAT, UMR 5245 CNRS-INP-UPS, Avenue de l'Agrobiopole, BP 32607, Auzeville tolosane, 31326 Castanet-Tolosan, France

B. Pourrut
LGCgE, Equipe Sols et Environnement, ISA, 48 boulevard Vauban, 59046 Lille Cedex, France

C. Dumat • E. Pinelli (✉)
INP-ENSAT, Université de Toulouse, Av. de l'Agrobiopôle, 31326 Castanet-Tolosan, France

EcoLab (Laboratoire d'écologie fonctionnelle), INP-ENSAT, UMR 5245 CNRS-INP-UPS, Avenue de l'Agrobiopole, BP 32607, Auzeville tolosane, 31326 Castanet-Tolosan, France
e-mail: pinelli@ensat.fr

M. Nadeem • M. Aslam
Department of Environmental Sciences, COMSATS Institute of Information Technology, Vehari 61100, Pakistan

6	Plant Heavy-Metal Tolerance Mechanisms	14
	6.1 Primary Heavy-Metal Tolerance Mechanisms	15
	6.2 Secondary Heavy-Metal Tolerance Mechanisms	16
	6.3 Glutathionylation	18
	6.4 Nitrogen Metabolism	20
	6.5 Antioxidant Enzymes	21
7	Conclusions and Perspectives	24
8	Summary	25
References		26

1   Introduction

Environmental contamination by hazardous environmental pollutants is a widespread and increasingly serious problem confronting society, scientists, and regulators worldwide (Debenest et al. 2010; Hajeb et al. 2011; Nanthi and Bolan 2012; Shahid et al. 2013a). Among these pollutants, the heavy metals, are a loosely-defined group of elements that are similar in that they all exhibit metallic properties, and have atomic masses >20 (excluding the alkali metals) and specific gravities >5 (Rascio and Navari-Izzo 2011). This group mainly includes transition metals, some metalloids, and the lanthanides and actinides. Heavy metals can be toxic to plants, animals and humans, even at very low concentrations. Heavy metals are natural components of the earth's crust and are present in different concentrations at different sites (Shahid et al. 2012a).

Heavy metal environmental pollution has occurred since ancient times, although their impact became worse during the industrial revolution from increased metal production and from development of new technologies that utilized these metals (Arshad et al. 2008; Nasim and Dhir 2010; Uzu et al. 2010; Vuai and Tokuyama 2011; Pourrut et al. 2011a, 2013; Bai et al. 2011; Tak et al. 2013; Shahid et al. 2013b) (Fig. 1). Unlike organic chemicals, the majority of heavy metals cannot be easily metabolized into less toxic compounds. These metals have long residence times in soils (Radwan et al. 2010; Ahmad and Ashraf 2011; Shahid et al. 2012b), and may continue to exert harmful effects on the environment over long periods

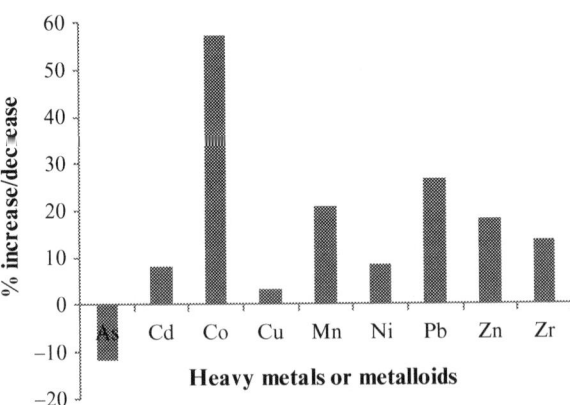

Fig. 1 Percent increase or decrease in annual production of heavy metals and metalloids during the period 2007–2011 [obtained from USGS (2012)]

(Giaccio et al. 2012), thereby representing a potential continuing threat to humans (Kerin and Lin 2010; Uzu et al. 2011a, b; Luo et al. 2012; Zhao et al. 2012; Foucault et al. 2013) and the environment (Schreck et al. 2011; Hunt et al. 2012).

The chemical, biological and physiological effects of heavy metal exposure to plants are of growing concern, because of their potential to accumulate therein and ultimately enter the food chain (Whiteside et al. 2010; Sarma et al. 2011; An et al. 2012; Schreck et al. 2012). The toxic impact of heavy metals on plants have been widely studied (Krzesłowska et al. 2010; Martínez-Fernández et al. 2011; Ahmad et al. 2011a; Evangelou et al. 2012; Hu et al. 2012; Shahid et al. 2013c), and different aspects thereon have been addressed in literature reviews (Pourrut et al. 2011b; Anjum et al. 2012).

Results of previous studies have shown that excessive accumulation of heavy metals in plant tissue can decrease root length, plant biomass, seed germination and chlorophyll biosynthesis (Singh et al. 2010). Inside the cell, heavy metals affect photosynthesis, respiration, mineral nutrition, enzymatic reactions and many other physiological factors (Pourrut et al. 2011b). A rather frequent and common effect of heavy metal toxicity in plants is increased production of reactive oxygen species (ROS). The production of ROS results from the interaction of heavy metals with electron transport activities, particularly in the chloroplast and mitochondrial membranes. The increased production of ROS can disrupt the redox status of cells, resulting in oxidative stress to exposed cells, leading to membrane dismantling, biological macromolecule deterioration, ion leakage, lipid peroxidation and DNA-strand cleavage (He et al. 2011; Carrasco-Gil et al. 2012; Chen et al. 2012). However, the toxic effects of heavy-metal-induced ROS on plant macromolecules vary and depend on the duration of exposure, stage of plant development, concentration of heavy metals tested, intensity of plant stress and the particular organs studied.

To prevent heavy-metal-induced ROS injuries, plants have developed various defense mechanisms by which they can transform ROS into less-toxic products (Tang et al. 2010; Álvarez et al. 2012). These mechanisms include: prohibiting metal entrance into plants, increased root excretion of metals, limiting toxic metal accumulation in sensitive tissue, chelation by organic molecules, metal binding to the cell wall and sequestration in vacuoles. These mechanisms help plants to sustain their cellular redox state and mitigate the damage caused by oxidative stress (Tang et al. 2010). The majority of these defense mechanisms depend on metabolic mediation of natural compounds such as phytochelatins (PCs), reduced glutathione (GSH), carotenoids and tocopherols, and enzymatic antioxidant systems including catalase (CAT and EC 1.11.1.6), superoxide dismutases (SOD and EC 1.15.1.1), ascorbate peroxidase (APX, EC 1.11.1.11), peroxidase (POD, EC 1.11.1.7), guaiacol peroxidase (GPX, EC 1.11.1.7), glutathione reductase (GR, EC 1.6.4.2), monodehydroascorbate reductase (MDHAR, EC 1.6.5.4) and dehydroascorbate reductase (DHAR, EC 1.8.5.1). The increased levels of these metabolic intermediary compounds and of antioxidant enzymes lead to increased stress tolerance against heavy-metal-induced ROS (He et al. 2011).

Considerable progress has been made in recent years in understanding how different plants respond physiologically to heavy-metal- and metalloid-induced stress. Despite this progress, information is limited on how these plant traits are regulated or are induced. How plants respond physiologically to heavy-metal-induced stress varies with plant species, metal type and species, and exposure conditions. Additionally, the mechanisms by which heavy metals induce oxidative stress and the different ways in which plants may respond to ROS are not completely elucidated. Therefore, predicting when, or how much heavy-metal-induced ROS production will occur, and how plants will detoxify these ROS are very important steps for improving our ability to assess risks or improve phytoremediation performance. With this in mind, it is our objective in this literature review to summarize key aspects of how plants are affected by heavy-metal-induced ROS production. In particular, we address (1) how plant exposure to heavy metals generates ROS, (2) what the toxic effects of ROS are to plant macromolecules such as DNA, proteins, carbohydrates and lipids, and (3) how plants defend themselves against, and eliminate ROS by enzymatic and non-enzymatic mechanisms.

2 What Are ROS?

"Reactive oxygen species" are generally regarded to exist when the following are present: (1) oxygen-derived free radicals such as hydroxyl (HO$^•$), superoxide anion ($O_2^{•-}$), peroxyl ($RO_2^•$), and alkoxyl (RO$^•$) radicals, or (2) oxygen-derived nonradical species such as hydrogen peroxide (H_2O_2), organic hydroperoxide (ROOH) and singlet oxygen ($^1/_2O_2$) (Corpas et al. 2011; Circu and Aw 2010). Although all of these oxygen-based toxic species are ROS, all ROS are not oxygen radicals. ROS are basically short lived, unstable and chemically very reactive molecules, possessing unpaired valence shell electrons (Wang et al. 2010).

3 ROS Production in Plant Metabolism

3.1 Natural Production of ROS in Plants

Under aerobic conditions, the generation of ROS is an inevitable aspect of life (Jaspers and Kangasjärvi 2010; Kovacic and Somanathan 2010; Swanson and Gilroy 2010; Wei et al. 2011; Foyer and Noctor 2012). Plant organelles such as mitochondria, chloroplasts and peroxisomes are considered to be major sources of ROS production in plant cells (Karuppanapandian et al. 2011a; del Río 2011; Borisova et al. 2012; Minibayeva et al. 2012; Pucciariello et al. 2012). In sun- or artificial lighting conditions, peroxisomes and chloroplasts are the main sources of ROS (Foyer and Noctor 2003). However, in darkness, plant mitochondria are considered to be the main site of ROS production (Foyer and Noctor 2003). The main sites of ROS production are the complex I and the complex III of the mitochondrial electron transport chain (Barranco-Medina et al. 2007). It is believed that almost 2% of the O_2 consumed by mitochondria is used to generate H_2O_2 (Becana et al. 2000). In the apoplast, ROS are produced as a consequence of NADPH oxidase activity (Achard et al. 2008; Weyemi and Dupuy 2012; Potocký et al. 2012).

During non-stressed cellular metabolism, O_2 is reduced to H_2O. During this process, ROS such as $O_2^{•-}$, H_2O_2 and $OH^•$ are produced as by-products, either by electron transfer or energy transfer reactions (Pucciariello et al. 2012; Borisova et al. 2012). The single electron reduction of O_2 generates the anion superoxide ($O_2^{•-}$). Superoxide is believed to be the precursor of most ROS and acts as a mediator in oxidative chain reactions. This anion is short-lived, which is easily dismutated to H_2O_2. In contrast to $O_2^{•-}$, H_2O_2 is highly stable and diffusible and is capable of inactivating cell molecules, even at a very low concentration. The main threat imposed by $O_2^{•-}$ and H_2O_2 lies in their ability to generate highly reactive $OH^•$ radicals (Møller et al. 2007; Bhatt and Tripathi 2011). In the presence of Fe, H_2O_2 and $O_2^{•-}$ interact in a Haber–Weiss reaction, which produces $OH^•$ (Minibayeva et al. 2012). The $OH^•$ is considered to be the most reactive ROS, owing to its ability to start radical chain reactions, which are considered to be responsible for producing toxic effects in plants (Mittler et al. 2004; Jones et al. 2011). Under normal conditions, an optimal ROS level is maintained by antioxidant enzymes.

3.2 Heavy-Metal-Induced Production of ROS in Plants

When exposed to heavy metals, plants are known to produce increased quantities of ROS (Table 1). This phenomenon is regarded to be among the earliest of biochemical changes, when plants are subjected to heavy metals stress (Jasinski et al. 2008; Yadav 2010; Grover et al. 2010; Lushchak 2011; Opdenakker et al. 2012). A serious imbalance occurs from the production and elimination of ROS, and this imbalance leads to dramatic physiological challenges to the plant that we call "oxidative stress" (Morina et al. 2010; Kováčik et al. 2010). Metals, such as Cu, Fe, Pb, Cd, Cr, As, Hg, Cr and Zn, all have the ability to induce the formation of ROS (Duquesnoy et al. 2010; Vanhoudt et al. 2010a, b; Márquez-García et al. 2011; Körpe and Aras 2011).

However, the phenomenon of ROS production is different for redox-active and redox-inactive metals (Pourrut et al. 2008; Opdenakker et al. 2012). Redox-active metals such as Fe and Cu catalyze Haber–Weiss/Fenton reactions:

$$\left(Cu^+ \rightleftharpoons Cu^{2+} + e^- \text{ and } Fe^{2+} \rightleftharpoons Fe^{3+} + e^-\right),$$

in which H_2O_2 is broken down into $OH^•$ at neutral pH (Valko et al. 2006; Sahi and Sharma 2005) (Fig. 2). In contrast, redox-inactive metals, such as Pb, Cd, As, Hg, Ni and Zn inhibit enzymatic activities as a result of their affinity for –SH groups on the enzyme (Mishra et al. 2006; Cuypers et al. 2011; Pourrut et al. 2011b). Redox-inactive metals form covalent bonds with protein sulfhydryl groups because of their electron-sharing affinities. Inactivation of enzymes results from the interaction of heavy metals with proteins, either at the catalytic site or elsewhere. Heavy metals, especially Pb, can also inactivate enzymes by binding to functional groups (COOH) present in proteins (Gupta et al. 2009, 2010). Moreover, displacement of essential cations by heavy metals from specific enzyme binding sites disrupts the ROS balance in cells, and results in ROS overproduction. For example, Zn, which acts as co-factor for many enzymes, can be replaced by heavy metals, causing enzyme

Table 1 Heavy-metal-induced reactive oxygen species (ROS) production in different plant species

Heavy metals	ROS	Plant species	References
Al	OH·, H_2O_2, O_2^-	Hordeum vulgare	Achary et al. (2012)
	NO·	Secale cereale	He et al. 2012)
		Triticum aestivum	
As	NOO·, H_2O_2, O_2^-	Oryza sativa	Singh et al. (2009)
Cd	H_2O_2	Arabidopsis thaliana	Martínez-Peñalver et al. (2012)
	H_2O_2	Chlorella vulgaris	Piotrowska-Niczyporuk et al. (2012)
	H_2O_2, O_2^-	Gracilaria dura	Kumar et al. (2012)
	H_2O_2	Brassica juncea	Ahmad et al. (2011b)
	H_2O_2	Medicago sativa	Antolín et al. (2010)
	H_2O_2	Ipomoea batatas	Kim et al. (2010)
	OH·, H_2O_2, O_2^-	Alocasia macrorrhiza	Liu et al. (2010a)
	H_2O_2, O_2^-	Solanum nigrum	Deng et al. (2010)
	H_2O_2	Brassica juncea	Guan et al. (2009)
	NO·, H_2O_2, O_2^-	Pisum sativum	Rodríguez-Serrano et al. (2009)
	OH·, H_2O_2, O_2^-	Ceratophyllum demersum	Aravind et al. (2009)
	H_2O_2	Triticum aestivum	Singh et al. (2008)
	H_2O_2, O_2^-	Arabis paniculata	Qiu et al. (2008)
	NO·	Triticum aestivum	Groppa et al. (2008)
	H_2O_2	Vicia faba	Lin et al. (2007)
	O_2^-	Mytilus galloprovincialis	Koutsogiannaki et al. (2006)
	H_2O_2	Nicotiana tabacum	Olmos et al. (2003)
	O_2^-	Lupinus luteus	Kopyra and Gwóźdź (2003)
	H_2O_2	Pisum sativum	Romero-Puertas et al. (2002)
	$O_2^{·-}$	Oryza sativa	Shah et al. (2001)
Cu	H_2O_2	Pisum sativum	Turchi et al. (2012)
	H_2O_2	Arabidopsis thaliana	Martínez-Peñalver et al. (2012)
	H_2O_2, O_2^-	Matricaria chamomilla	Kováčik et al. (2010)
	H_2O_2	Ipomoea batatas	Kim et al. (2010)
	H_2O_2	Medicago sativa	Antolín et al. (2010)
	NO·, H_2O_2	Lycopersicon lycopersicum	Wang et al. (2010)
	H_2O_2, O_2^-	Withania somnifera	Khatun et al. (2008)
Mn	H_2O_2, O_2^-	Cucumis sativus	Shi and Zhu (2008)
Ni	H_2O_2, O_2^-	Hypnum plumaeforme	Sun et al. (2010)
		Thuidium cymbifolium	
		Brachythecium piligerum	
Pb	H_2O_2	Vicia faba	Shahid et al. (2012a, b, c, d)
	H_2O_2	Chlorella vulgaris	Piotrowska-Niczyporuk et al. (2012)
	H_2O_2, O_2^-	Vallisneria natans	Wang et al. (2010)
	O_2^-	Spinacia oleracea	Wang et al. (2010)
	H_2O_2	Triticum aestivum	Yang et al. (2010)
	H_2O_2, O_2^-	Hypnum plumaeforme	Sun et al. (2010)
		Thuidium cymbifolium	
		Brachythecium piligerum	
	OH·, H_2O_2, O_2^-	Alocasia macrorrhiza	Liu et al. (2010a)
	H_2O_2	Medicago sativa	Antolín et al. (2010)
	H_2O_2	Sedum alfredii	Liu et al. (2008)
	O_2^-	Vicia faba	Pourrut et al. (2008)
	H_2O_2	Elsholtzia argyi	Islam et al. (2008)
	H_2O_2, O_2^-	Sedum alfredii	Huang et al. (2008)
	H_2O_2	Oryza sativa	Chen et al. (2007)
	O_2^-	Lupinus luteus	Kopyra and Gwóźdź (2003)
Zn	H_2O_2	Pisum sativum	Turchi et al. (2012)
	H_2O_2	Ipomoea batatas	Kim et al. (2010)
	O_2^-	Mytilus galloprovincialis	Koutsogiannaki et al. (2006)

$O_2^{·-}$, superoxide anion; HO·, hydroxyl; H_2O_2, hydrogen peroxide; NO·, nitric oxide; NOO·, nitrogen dioxide

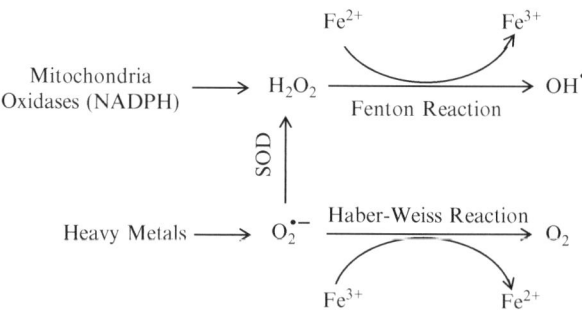

Fig. 2 The Haber–Weiss and Fenton reaction pathways; SOD= Superoxide Dismutase [modified from Kehrer (2000)]

inhibition and oxidative stress. Heavy metals are also capable of depleting GSH inside plant cells (Pourrut et al. 2011b, 2013; Lee et al. 2012). When this happens, heavy metals deplete the major antioxidants that exist within cells, which disrupts the ROS balance. Heavy metals also enhance ROS production via binding and consuming GSH and its derivatives directly, which are required to scavenge any ROS generated (Lee et al. 2003). In addition, plasma-membrane-bound NADPH oxidase is involved in heavy-metal-induced oxidative stress (Sagi and Fluhr 2006; Pourrut et al. 2008, 2013; Weyemi and Dupuy 2012; Potocký et al. 2012). Plasma membrane-bound NADPH oxidases can utilize cytosolic NADPH to generate $O_2^{\cdot -}$, which is quickly dismutated to H_2O_2 by SOD (Pourrut et al. 2008). The ROS formed by the NADPH oxidase exists outside the plasma membrane, where the pH is normally lower than that inside the cell (Sagi and Fluhr 2006). Heavy-metal-induced ROS generation via NADPH oxidase was reported in Cd-treated *Pisum sativum* (Rodríguez-Serrano et al. 2006), Ni-treated *Triticum durum* (Hao et al. 2006) and Pb-treated *Vicia faba* (Pourrut et al. 2008). Moreover, Ca^{2+} and protein kinases have also been reported to have a role in heavy-metal-induced ROS production by activating NADPH oxidase (Yeh et al. 2007; Sahi and Sharma 2005; Pourrut et al. 2013).

4 Roles of ROS in Plant Metabolism

Traditionally ROS were considered to be toxic by-products of aerobic metabolism, but several recent reports clarified the essential roles of ROS in living organisms (Bailly et al. 2008; Rai et al. 2011; Bartoli et al. 2012; Swanson et al. 2011). These essential roles include:

- Plant metabolic defense under stress (Juan et al. 2010; Shin et al. 2011; Rai et al. 2011; Gémes et al. 2011),
- Plant disease resistance (i.e., bacterial and viral) (Jaspers and Kangasjärvi 2010; Shin et al. 2011; Kranner et al. 2010; Rai et al. 2011),

- Plant signal transduction that controls programmed cell death (Pitzschke and Hirt 2006; Blokhina and Fagerstedt 2010; Gill and Tuteja 2010; Rai et al. 2011; Corpas et al. 2011),
- Plant growth regulation (e.g., cell wall loosening) (Kranner et al. 2010; Šírová et al. 2011; Arasimowicz-Jelonek et al. 2011),
- Regulation of photorespiration and photosynthesis (Edreva 2005; Gill and Tuteja 2010),
- Initiating mitogen-activated protein kinase cascades (Jaspers and Kangasjärvi 2010),
- Regulation of root physiology (root hair development, root cell wall loosening and stiffening) (Foreman et al. 2003),
- Regulation of stomatal movement (Yu et al. 2009; Gill and Tuteja 2010),
- Regulation of the cell cycle (Mittler et al. 2004; Gadjev et al. 2008; Gill and Tuteja 2010),
- Fruit ripening and senescence (Karuppanapandian et al. 2011a, b), and
- Alleviation of seed dormancy (Oracz et al. 2009; Kranner et al. 2010; Whitaker et al. 2010; Roach et al. 2010).

The role of H_2O_2 as a signaling molecule, when it intervenes to defend against heavy metal stress has gained considerable attention in recent years. H_2O_2 can mediate the activities of protein kinases, protein phosphatases and transcription factors (Opdenakker et al. 2012). Protein kinases can regulate gene transcription by repressing or activating transcription factors (Pandey and Somssich 2009). Several authors have reported that ROS and protein kinases are activated, in response to heavy metal exposure. Yeh et al. (2007) reported the induction of kinases via ROS production from Cu^{2+} and Cd^{2+} stress. Moreover, cadmium exposure is reported to have induced protein kinase transcripts via the accumulation of ROS in *Zea mays* (Wang et al. 2010) and *Arabidopsis thaliana* (Liu et al. 2010). However, very little is known about the mechanisms and the exact signaling pathways that operate behind these processes in plants that are under heavy metal stress.

5 Toxic Effects of Heavy-Metal-Induced ROS on Macromolecules in Plants

Heavy-metal-induced ROS can elicit widespread damage to plants, examples of which are enzyme inhibition, protein oxidation, lipid peroxidation and DNA and RNA damage (Martínez Domínguez et al. 2009; Cuypers et al. 2011). It has been reported that the indirect effect of heavy metals on plants macromolecules via ROS production is more toxic and rapid than the direct effect (Pourrut et al. 2011b). Reactive oxygen species are involved in the early steps of heavy-metal-induced toxicity to plants, and hence act as initiators of heavy metal toxicity (Shahid et al. 2012c; Martínez-Peñalver et al. 2012).

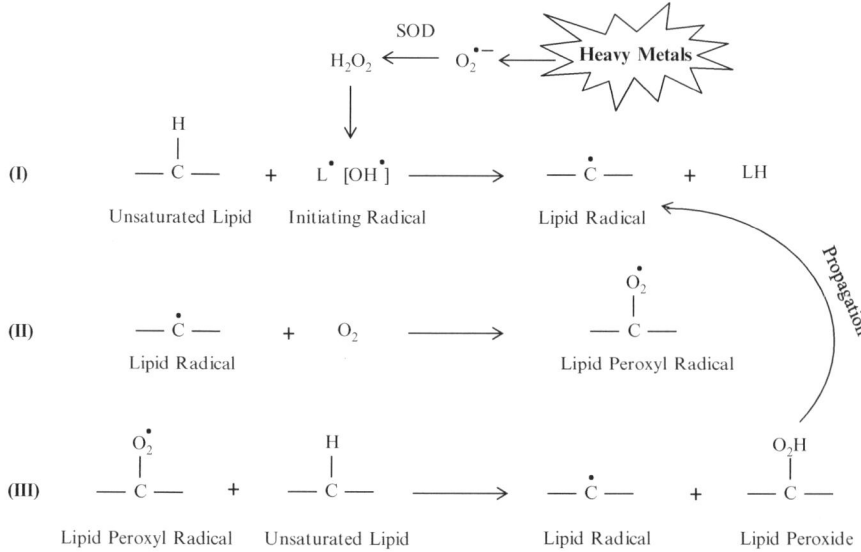

Fig. 3 Depictions of the possible mechanisms by which metals induce lipid peroxidation. The mechanism of heavy-metal-induced lipid peroxidation is initiated most likely via OH·. The process involves three distinct stages: initiation, progression and termination [modified from Bhattacharjee (2005)]

5.1 Lipid Peroxidation

Lipids are very important cellular components that play vital roles in various biological processes, such as providing energy for cellular metabolism, building cell membranes, and maintaining organelle and cell integrity and composition (Wallis and Browse 2002; Xiao and Chye 2011). Inside the plant, plasma cell membranes are the primary target of heavy metal action (Cuypers et al. 2011). Heavy metals are known to cause lipid peroxidation via ROS production (Fig. 3) (Cuypers et al. 2011; Wahsha et al. 2012; Márquez-García et al. 2012; Chen et al. 2012). Lipid peroxidation causes deterioration of cell membranes, and is one of the most harmful effects induced in plants by heavy-metal exposure (Pourrut et al. 2013). Lipid peroxidation may result from increased lipoxygenase activity, which initiates the formation of oxylipins (Porta and Rocha-Sosa 2002). Lipoxygenase has been reported to play an important role in heavy-metal-induced oxidative stress in *Gracilaria dura*, *Lessonia nigrescens* and *Arabidopsis thaliana* (Smeets et al. 2008; Kumar et al. 2012; Vanhoudt et al. 2011).

The phenomenon of lipid peroxidation is most common in polyunsaturated fatty acids and involves three distinct stages: initiation, progression and termination (Pourrut et al. 2011b; Bhattacharjee 2012). Reactive oxygen species are the most common initiators of lipid peroxidation in living cells. These ROS remove the hydrogen atom from a methylene group ($-CH_2-$), thus, giving rise to peroxyl radicals (Grover et al. 2010; Singh et al. 2010). The ROS-induced initiation of lipid peroxidation varies with stress condition and cell type. Under normal conditions, lipid peroxidation in green plant tissues is generally initiated by $O_2^{\cdot-}$, a non-radical electrophilic by-product of light capture in photosystem II (PSII) (Triantaphylidès and Havaux 2009). Heavy metals are known to inhibit PSII, and thus increase $O_2^{\cdot-}$ production in leaves, which leads to increased lipid peroxidation (Triantaphylidès et al. 2008; Triantaphylidès and Havaux 2009; Farmer and Mueller 2013). In chlorophyll-lacking tissues, lipid peroxidation is started by $OH^{\cdot}$, a radical produced by Fe- or Cu-catalysed degradation of H_2O_2 (Farmer and Mueller 2013). Although $O_2^{\cdot-}$ and H_2O_2 are capable of initiating the reactions that are responsible for lipid peroxidation, only $OH^{\cdot}$ is sufficiently reactive, especially in the presence of transition metals such as Cu or Fe (Bhattacharjee 2005; Pourrut et al. 2013). One electron redox cycle results in the formation of peroxyl and alkoxyl radicals (Karuppanapandian et al. 2011a). The fatty acid radical formed is not very stable. In an aerobic environment, oxygen reacts with the fatty acid, thereby creating another unstable peroxyl-fatty acid radical. Once initiated, $ROO^{\cdot}$ groups are capable to continue the peroxidation chain reaction by receiving a hydrogen atom from neighbouring polyunsaturated fatty acids (Bhattacharjee 2005; Karuppanapandian et al. 2011a). The resulting lipid hydroperoxide is a highly unstable molecule and decays into several reactive species such as lipid epoxides, aldehydes (malonyldialdehyde), lipid alkoxyl radicals, alkanes and alcohols (Bhattacharjee 2005). The cycle continues from the presence of fatty acid side chains that are in close proximity to plant membranes, which facilitates autocatalytic propagation of lipid peroxidation.

Generally lipid peroxidation causes: (1) increased membrane leakiness to substances that do not normally cross membranes, other than via specific channels, (2) decreased membrane fluidity, which makes it easier for phospholipids to be exchanged between the two halves of the bilayer, and (3) damage to membrane proteins that inactivate receptors, enzymes, and ion channels. Several studies revealed toxic effects from lipid peroxidation in plants (Yamauchi and Sugimoto 2010; Farmer and Mueller 2013). Some recent studies reported that heavy metal toxicity to different physiological processes occurs via ROS-induced lipid peroxidation (Shahid et al. 2013d). The by-products of lipid peroxidation also strongly affect photosynthetic reactions. For example, acrolein, linolenic acid-13-ketotriene and 12-oxo-phytodienoic acid are well known to induce toxic effects on PSII (Alméras et al. 2003). Exogenous acrolein is reported to deplete chloroplast glutathione pools (Mano 2012). Lipid peroxidation also causes covalent modification of plant proteins due to the binding of electrophilic lipid fragments with proteins (Farmer and Mueller 2013). This covalent binding occurs when nucleophilic atoms (e.g., S or N) bind to the β-carbon of α,β-unsaturated carbonyl groups. Nowadays, increased attention is being given to the damaging effects of lipid peroxidation products, which can be monitored by using of transgenic approaches (Mano 2012).

5.2 DNA Damage

Heavy-metal-induced genotoxicity in plant cells is a complex phenomenon, and the mechanisms behind this process are not yet well understood (Aina et al. 2004; Tuteja et al. 2009; Cuypers et al. 2011; Zhu et al. 2011; Shen et al. 2012). According to some authors, heavy-metal-induced DNA damage is not direct but occurs indirectly through ROS production (Gichner et al. 2006; Gupta and Sarin 2009; Barbosa et al. 2010; Hirata et al. 2010, 2011). Heavy-metal-induced DNA damage has been reported in several plants, examples of which are, *Trifolium repens* (Aina et al. 2004), *Cannabis sativa* (Aina et al. 2004), *Allium cepa* (Barbosa et al. 2010), *Vicia faba* (Marcato-Romain et al. 2009a; Pourrut et al. 2011c), *Boletus edulis* (Collin-Hansen et al. 2005), and *Nicotiana tabacum* and *Solanum tuberosum* (Gichner et al. 2006).

Among ROS, OH$^\bullet$ is the most reactive entity in damaging all components of the DNA molecule (Jones et al. 2011). Reactive oxygen species interactions with DNA results in: damage to cross-links, base deletions, base modifications, strand breaks and damage to pyrimidine dimers (Tuteja et al. 2001; Gastaldo et al. 2008). Among these affected DNA sites, base deletion is the most frequent DNA damage induced by either heavy metals, ionizing radiation or ultra violet radiation (Gastaldo et al. 2008). DNA has four different potential sites to which metals may bind, i.e., the ribose hydroxyls, the negatively charged phosphate oxygen atoms, the exocyclic base keto groups and the base ring nitrogens (Oliveira et al. 2008). Most transition metal ions interact in a complex way with DNA: more than two different sites are generally involved. Heavy metals generally bind directly to the bases, with the N7 atom of purines or N3 of pyrimidines and indirectly to the phosphate groups (Anastassopoulou 2003). In vitro studies indicated that heavy metals like Cd, Cr, Cu, Hg, Pb and Zn interact with DNA, particularly at sulfhydryl groups and the phosphate backbone (Sheng et al. 2008). Moreover, heavy metals may alter gene expression (Rossman 2000) and they appear to interact with Zn-fingers, which bind tetrahedrally to cysteine (Cys) thiolates and/or histidine imidazole groups to maintain the DNA three-dimensional structure (Witkiewicz-Kucharczyk and Bal 2006). DNA damage can occur either from replication errors, induction of signal transduction pathways, induction of transcription, cell membrane destruction and/or genomic instability (Cooke et al. 2003). In plants and other living organisms, damage inflicted on DNA and repair mechanisms generally occur concomitantly, making these processes both complex and difficult to independently assess (Gastaldo et al. 2008).

When ROS interact with DNA, oxidized bases are frequently generated (Hirano and Tamae 2010). Among the different forms of oxidative DNA damage, effects with 8-oxoguanine has been most extensively investigated (Hirano and Tamae 2010), and is also an event that may lead to neoplastic transformation (Bal and Kasprzak 2002). Using a plasmid-relaxation assay, Yang et al. (1999) demonstrated that Pb and Cd promoted DNA strand-breakage and formed 8-hydroxydeoxyguanosine (8-OHdG) adducts in DNA. Recently, Hirata et al. (2011) showed As- and Cr-induced translesion DNA synthesis due to their increased affinity for DNA containing 8-OHdG.

Heavy-metal-induced damage to DNA may also result in the production of micronuclei, which produce chromosome breaks or mitotic anomalies that require passage through mitosis to be recognisable (Marcato-Romain et al. 2009b). According to Johnson (1998), heavy metals are capable of interfering with the spindle apparatus of dividing cells to produce DNA damage. Cenkci et al. (2009) described Pb-induced genotoxicity, using a random amplified polymorphic DNA (RAPD) profile, in *Brassica rapa* exposed to 0.5 to 5 mM concentrations of lead nitrate. Radić et al. (2011) demonstrated damage to DNA (estimated by tail extent moment) in *Lemna minuta* exposed to heavy metals from industrial wastewater. Recently, Shahid et al. (2011) reported the Pb-induced production of micronuclei in *Vicia faba* root tips via ROS production. More recently, Pourrut et al. (2011b) demonstrated a close link between oxidative stress induced by Pb, DNA strand breaks and micronuclei formation in *Vicia faba* root tips.

5.3 Protein Damage

Heavy metals may also cause toxic effects in the structure of plant proteins (Tan et al. 2010; Luque-Garcia et al. 2011). Protein synthesis is the primary target of ROS damage in plants (Nishiyama et al. 2011). This heavy-metal-induced change in protein quantity or quality can occur via several mechanisms, e.g., binding of the metal ions to free thiols and other functional groups of proteins, replacement of Zn and other essential metal ions by free heavy metal ions in metal-dependent proteins, etc. Whatever the location of heavy metal-induced ROS, they generally interact with proteins that contain sulfur-containing amino acids and thiol groups. Proteins are more susceptible to heavy metal ions during the process of folding, than are proteins that have already reached their native state (Sharma et al. 2008).

Heavy-metal-induced ROS also cause a quantitative reduction in total protein content of cells (Mishra et al. ; Garcia et al. 2006). This quantitative decrease in total protein content results from various heavy metals effects: they modify gene expression (Kovalchuk et al. 2005), increase ribonuclease activity (Gopal and Rizvi 2008), consume amino acids to scavenge ROS (Gupta and Sinha 2009), and reduce free amino acid content (Gupta et al. 2009) that is linked to alteration in nitrogen metabolism (Chatterjee et al. 2004). Heavy metal ions form complexes with proteins by binding with –COOH, –NH$_2$ and –SH groups (Tan et al. 2010). As a result, these modified biological molecules cannot function properly as a result of their structural modification, and this produces cell malfunction. When heavy metals bind to these active groups of proteins, they inactivate different enzyme systems, or alter protein structure, which is related to the catalytic properties of enzymes. Reactive oxygen species do oxidize the following protein amino acid side groups: Cys, Met, His, Arg, Lys, Pro, Tyr and Trp. Cadmium treatment raised the carbonylation level from 4 to 5.6 nmol/mg protein in *Pisum sativum* plants (Romero-Puertas et al. 2002). Most of these reactions are irreversible, although in the specific case of thiol-group oxidation, enzyme-catalyzed re-reduction is possible (Rouhier et al. 2006).

Recent findings suggest that protein oxidation events are most likely to occur in proteins that are extremely close to the site of ROS production. Certain metal ion co-factors, such as Fe–S, are particularly susceptible to oxidation. Heavy metal exposure to plants not only causes a quantitative change to protein content, but also may alter the qualitative composition of cell proteins. The protein composition of root cells in *V. faba* seedlings was altered when exposed to Pb (Beltagi 2005), and this can result from the modification in transcriptome profile of numerous enzymes such as: cysteine proteinase, isocitrate lyase, arginine decarboxylase and serine hydroxymethyltransferase (Kovalchuk et al. 2005).

Heavy metals also may produce indirect effects on protein functioning that curtails protein synthesis or inhibits protein functioning (Pena et al. 2008). For example, the plant proteolysis system helps to regulate protein processing and intracellular protein levels, and removes abnormal or damaged proteins from the cell (Buchanan et al. 2000). The proteolytic system is mainly localized inside certain organelles, e.g., cytoplasm and the nucleus (Rawlings 2004). Cadmium has been reported to cause oxidation of the proteasome in *Zea mays* (Pena et al. 2007) and *Helianthus annuus* plants (Pena et al. 2006). This enhancement of the proteasome activity prevents accumulation of oxidatively damaged proteins in the cell (Pena et al. 2007).

5.4 Damage to Plant Carbohydrates

Carbohydrates are ubiquitous energy sources, and are key macromolecules for their role in plant metabolism and structure (Guan-fu 2011; Dong et al. 2011). Carbohydrates are the major products of photosynthesis and act as transport molecules in plant growth, development and storage (Couée et al. 2006). They are involved in response mechanisms to different stressors, osmotic adjustment, and nutrient and metabolic signaling molecules (Hummel et al. 2009). They also help to maintain plasma membrane integrity (Guan-fu 2011), feed the NADPH-producing metabolic pathways involved in ROS scavenging, and interact with plant hormone signaling through molecules such as the auxins and cytokinins (Rolland et al. 2002), gibberellin, abscisic acid and ethylene (Price et al. 2004). Heavy metals are known to affect plant sugar content through ROS-induced oxidative stress. Interaction between soluble sugar content and ROS cause pollen abortion in *Triticum aestivum* (Lehner et al. 2008) or decreased pollen viability in *Oryza sativa* (Guan-fu 2011), which might be due to the interplay between programmed cell death and ROS. Any expression of sugar transporter genes that are induced by heavy metal stress may reduce the oxidant damage caused by overproduction of ROS (Nguyen et al. 2010). Glucose is reported to enhance cellular defences against cytotoxicity of H_2O_2 in plants, and enhances plantlet survival (Averill-Bates and Przybytkowski 1994). Under intense oxidative stress conditions, ROS affects the structure of carbohydrates (Zadák et al. 2009). When thus affected, plant defense mechanisms are weakened and plant macromolecules (including glucose) become vulnerable to heavy metal toxicity.

5.5 Interference with Signalling

Heavy metals interfere with cell signalling via mechanisms that are poorly understood. Effects of heavy metals on cell signalling may be direct as a result of the interaction of metals with proteins, or indirect from the formation of metal-induced ROS. It has been proposed that heavy-metal-induced disregulation of signalling events play a key role in the response of heavy metal toxicity as well as in damage development. Metals affect the gene expression, transcription and activation of numerous signalling proteins, including growth factor receptors, G-proteins and tyrosine kinases (Harris and Shi 2003). In plants, several studies have shown that heavy metals (Cu, Zn, Pb and Cd) intervene with mitogen kinase signalling cascades. Mitogen-activated protein kinase (MAPK) pathways incorporate various signalling stimuli, and specific elements are also activated by ROS (Zhang and Klessig 2001). These MAPKs are rapidly activated in *Medicago sativa* by an excess of Cu (Jonak et al. 2004). However, Cd exposure activates MAPKs in *Medicago sativa* after a considerable delay (Jonak et al. 2004). The titer of jasmonic acid, salicylic acid and ethylene increases in plants after exposure to heavy metals (Pál et al. 2005), which then enhances H_2O_2 generation (Zawoznik et al. 2007) and interferes with cell signalling. Romero-Puertas et al. (2007) explained how the redox component scheme works, and explained how signalling molecules positively or negatively adjust the expression of antioxidant genes during long-term Cd stress in *Pisum sativum*.

6 Plant Heavy-Metal Tolerance Mechanisms

To survive, plants have to constantly cope with stress. Certain plants (especially heavy metal hyperaccumulator plants) operate well even under extreme ROS production situations that are caused by heavy metal toxicity. In fact, plants have evolved an array of defense mechanisms to combat oxidative damage, for the purpose of restricting cell injury and tissue dysfunction (Shulaev et al. 2008; Benekos et al. 2010; Ruan et al. 2011). Such defense mechanisms act separately or simultaneously in plants to scavenge any ROS over-production. However, what specific plant defense mechanism are active, and the efficiency of it, depends on the plant species, plant maturity, type of metal involved, and the level and duration of exposure.

Generally, stress-tolerant plants better defend themselves against ROS than do stress-susceptible species (Liu and Pang 2010). Hyperaccumulator plants are efficient at detoxifying and sequestering heavy metals, which enable them to accumulate high metal levels in their shoot tissues, without suffering phytotoxic effects (Rascio and Navari-Izzo 2011). Such preferential heavy metal detoxification/sequestration does occur in specific plant structures, such as the epidermis (Freeman et al. 2006), trichomes (Küpper et al. 2000) and even the cuticle (Robinson et al. 2003), where they cause toxicity to the photosynthetic apparatus, if not detoxified.

6.1 Primary Heavy-Metal Tolerance Mechanisms

Heavy metals mainly enter plants from soil through the roots (Uzu et al. 2009; Tang et al. 2010). Heavy metals, especially Pb, are adsorbed onto the root surface before uptake and become bound to carboxyl groups of mucilage uronic acid or to the polysaccharides of the rhizoderm cell surface (Seregin et al. 2004; Pourrut et al. 2011b). Such binding of heavy metals to exchange sites at the root surface is a commonly employed plant strategy to limit heavy metal absorption into root cells; the entrapment occurs in the apoplast by binding the metals to exuded organic acids or anionic groups of cell walls (Jiang and Liu 2010). In response to heavy metal toxicity, root thickness can increase, and thereby increase the amount of metal adsorbed onto the root surface; when this occurs, the consequence is to reduce metal penetration into roots (Krzesłowska et al. 2009, 2010). Probst et al. (2009) observed increased cell wall thickness of *Vicia faba* as an ultrastructural alteration caused by a high metal level. Liu et al. (2004) and Andrade et al. (2004) reported similar increases in cell wall thickness, respectively, in shoots of *Vicia faba* that were exposed to Cu or Cd, and in marine macroalgus exposed to Cu. Such increases are believed to be associated with enhanced peroxidase activity (Liu et al. 2004; Probst et al. 2009). This enzyme catalyzes lignin synthesis (Arduini et al. 1995) and is generally produced in higher plants exposed to heavy metals (Prasad 1996). Probst et al. (2009) observed high amounts of electron-dense particles of metals (Pb and Zn) on the surface, and within the cell walls of *Vicia faba* roots. Similar Pb deposits were shown to exist along plasma membranes of *Sesbania* root cells by Sahi and Sharma (2005). Krzesłowska et al. (2009) reported reduced penetration of Pb into the plasma membrane in *Funaria hygrometrica* from increased cell wall thickness, as a result of Pb binding with JIM5-P, within the cell wall. However, Pb bound to JIM5-P can be remobilized by endocytosis (Krzesłowska et al. 2010). In has been reported in several studies that Pb is adsorbed onto roots in many plant species: *Vigna unguiculata* (Kopittke et al. 2007), *Brassica juncea* (Meyers et al. 2008), *Festuca rubra* (Ginn et al. 2008), *Lactuca sativa* (Uzu et al. 2009) and *Funaria hygrometrica* (Krzesłowska et al. 2010). The degree of adsorption of metals onto plant root surface varies with the physico-chemical properties of rhizosphere soil, and plant and metal type (Saifullah et al. 2009; Pourrut et al. 2011b). The adsorption of metals onto root surfaces reduces their entrance into plants, which is considered to be beneficial in the case of vegetables (Pourrut et al. 2011b).

Another defense mechanism plants adopt is to reduce the translocation of heavy metals to aerial plant parts. Most of the heavy metals absorbed by plants are sequestered in plant root cells. In root cells, toxic metals are detoxified by complexation with organic acids, amino acids or sequestered into vacuoles (Rascio and Navari-Izzo 2011; Pourrut et al. 2011b). Such complexation restricts the transfer of heavy metals towards aerial plant parts, thus protecting leaf tissues, and particularly the metabolically active photosynthetic cells from heavy metal damage (Rascio and Navari-Izzo 2011). Increased sequestration of heavy metals in root cells is achieved by several mechanisms: they precipitate as insoluble salts in intercellular spaces (Meyers et al. 2008), they are immobilized by negatively charged pectins within the cell wall (Arias et al. 2010), they accumulate in plasma membranes (Jiang and Liu 2010), or are sequestered in the vacuoles of rhizodermal and cortical cells (Kopittke et al. 2007). Many researchers have reported that >90% of heavy metals present accumulate in plant root cells of many plant species. Examples are: *Vigna unguiculata* (Kopittke et al. 2007), *Pisum sativum, Phaseolus vulgaris* and *Vicia faba* (Pourrut et al. 2011a), *Arabidopsis thaliana* (Vanhoudt et al. 2010a) *Avicennia marina* (Yan and Lo 2011), *Sedum alfredii* (Gupta et al. 2010), *Allium sativum* (Jiang and Liu 2010), *Lolium perenne* (Jia et al. 2011), *Oryza sativa* (Hu et al. 2011), *Erica andevalensis* (Mingorance et al. 2012) and *Chrysopogon zizanioides* (Danh et al. 2011). The phenomenon of increased amounts of metals being restricted to accumulating in roots is more common to Pb than to other heavy metals.

6.2 Secondary Heavy-Metal Tolerance Mechanisms

When plants take up high levels of heavy metals, toxicity is prevented only if the plants have a strong sink adequate for storing the toxic metals (Wojas et al. 2010; Hassan and Aarts 2011). By having such sinks, plants can evade the toxic effects of these metals. Vacuolar sequestration is an important feature that maintains plant metal homeostasis, and detoxifies heavy metals (Maestri et al. 2010). The hyperaccumulator plants have the ability to limit negative effects of metals by sequestering and/or binding them to molecules or plant structures. Heavy metals are detoxified in aerial parts of hyperaccumulators plants as a result of ligand binding or entrapment by vacuoles (Rascio and Navari-Izzo 2011). Vacuolar transporters partly fulfil this role, by contributing to the partitioning of metals into the vacuole (Martinoia et al. 2007).

The vacuole is the final destination for practically all toxic substances. There are several pathways by which metals are sequestered vacuoles. Genomic sequencing analysis has identified various families of transporters that are involved in heavy metal homeostasis in plants (Klatte et al. 2009; Chaffai and Koyama 2011). These transporter families include ATP-binding cassettes (ABC), heavy metal ATPases (HMAs), Zrt/Irt-like protein (ZIP), cation exchangers (CAXs), natural resistance-associated macrophage (NRAMP) and cation diffusion facilitators (CDF) (Grotz and Guerinot 2006; Hall and Williams 2003). Among these, CDF ABC and NRAMP have been identified as being critical for heavy metal tolerance (Hanikenne et al. 2005; Chaffai and Koyama 2011).

Metallothioneins (MTs) and phytochelatins are the best characterized and important metal-binding ligands in plant cells (Rea 2012). Phytochelatins are small, heavy-metal-binding polypeptides that have the general structure of (γ-Glu-Cys)nGly (n=2–11). Phytochelatins belong to different classes of cysteine-rich heavy metal-binding protein molecules. Heavy metals are capable of stimulating the production of PCs, and activating the enzyme phytochelatin synthase (PCS) (Vadas and Ahner 2009; Jiang and Liu 2010). The synthesis of PCs is catalyzed non-translationally by PCS, which is activated by metal ions such as Cd, Pb, Zn, and Cu (Andrade et al. 2010; Ogawa et al. 2011). In plants, these natural chelators bind and transport heavy metals to cell vacuoles (Israr et al. 2011). The transport of the metal-PC complex to vacuoles is thought to be facilitated by ABC transporters (Prévéral et al. 2009; Park et al. 2012), which for *Oryza sativa* seedlings, are encoded by OsPDR5/ABCG43 (Oda et al. 2011). PCs bind and transport heavy metals by forming mercaptide bonds with them (Verbruggen et al. 2009; Semane et al. 2010). Generally, PCs bind metals in the cytosol, and the resulting PC–metal complex is sequestrated in vacuoles (Ogawa et al. 2011), thereby reducing the concentration of free metal ions in the cytosol. In this way, these natural ligands inhibit ROS production that results from heavy metal interactions with the delicate redox system. In in-vivo studies, Yadav (2010) reported that PCs were involved in the cellular detoxification and accumulation of heavy metals as a result of their ability to form stable metal-PC complexes. Gisbert et al. (2003) reported that the induction and over-expression of a *Triticum aestivum* gene encoding phytochelatin synthase (TaPCS1) significantly increased uptake and tolerance of *Nicotiana glauca* to Pb and Cd.

Glutathione (GSH; γ-glutamatecysteine-glycine), a sulfur containing tri-peptide, is among the most important and critical of the low molecular weight biological thiols. Glutathione protects plants from heavy metal toxicity by quenching metal-induced ROS (Vanhoudt et al. 2010a; Seth 2012; Noctor et al. 2012). Glutathione reacts nonenzymatically with a series of ROS by forming thiyl radicals (Halliwell and Gutteridge 1999). Thiyl radicals may generate $O_2^{\cdot -}$, which can be neutralized by SOD/CAT enzymes. It is worth noting that GSH also reacts with the lipid peroxidation metabolite 4-hydroxy-2-nonenal (Wonisch et al. 1997), and plays a role in the initial resistance against malondialdehyde, another highly toxic lipid peroxidation product (Turton et al. 1997).

Moreover, it is a substrate for PC biosynthesis, and certain related proteins play a key role in detoxifying heavy metals (Huang and Wang 2010; Ogawa et al. 2011). It is noteworthy that metals do not directly activate PCS activity, but rather, a GSH-metal complex is formed, (i.e., in which the metal binds to a thiol group), which activates PCS (Na and Salt 2010). Glutathione synthesis is catalyzed by two ATP-dependent enzymes, γ-glutamylcysteine synthetase (GSH1) and glutathione synthetase (GSH2). Heavy metal exposure can induce different GSH genes, such as glutathione synthetase, glutamyl cysteine synthetase, glutathione peroxidase and glutathione reductase. A deficiency of GSH affects defense gene expression and the hypersensitive response in plants (Dubreuil-Maurizi et al. 2011). Glutathione is reported to enhance proline accumulation in heavy-metal-stressed plants, a role that is correlated with reduced damage to membranes and proteins (Liu et al. 2009). Generally, PCs and GSH are simultaneously stimulated in plants to detoxify heavy metals. However, Gupta et al. (2010) reported the induction of GSH alone for detoxification of heavy metals in *Sedum alfredii*. The enhanced production of GSH does not always increase plant tolerance or detoxify heavy metals to reduce plant stress (Xiang et al. 2001). Therefore, GSH alone may not be adequate to resist heavy-metal stress in plants (Noctor et al. 1998; Yadav 2010).

Glutathione also plays an important indirect role in detoxifying heavy metals via activating the PCS enzyme. Once sufficient GSH levels are achieved during heavy metal stress, PCS become active and catalyzes the formation of PC–metal complexes (Yadav 2010). PCS are activated when a heavy metal and two GSH molecules form a thiolate complex (Cd–GS2 or Zn–GS2). Activation of PCS also results in the transfer of one γ-Glu-Cys moiety to a free GSH molecule or to a previously synthesized PC (Singla-Pareek et al. 2006). Depletion of GSH may result from its consumption for PCs synthesis (Mishra et al. 2006), or from direct binding with heavy metal ions (Andra et al. 2009a, b).

6.3 Glutathionylation

The thiol group of the amino acid cysteine is extremely vulnerable to ROS (oxidative damage), due to its high sensitivity to oxidation. To protect proteins from oxidation, plant cells have developed a tolerance mechanism, glutathionylation, which results in a reversible posttranslational modification of protein thiols (Michelet et al. 2006; Zaffagnini et al. 2012a). During glutathionylation, the protein thiols are oxidized to various reversible products, such as S-glutathionylation, sulfenic or sulfinic acids, and intra- or inter-protein disulfide bonds (Li and Zachgo 2009). The reaction mechanism of glutathionylation involves an exchange of a thiol/disulfide between GSSG and a protein thiol as following:

$$\text{Protein-SH} + \text{GSSG} \rightleftharpoons \text{Protein-SSG} + \text{GSH}$$

Several proteomic studies have demonstrated the glutathionylation of a number of chloroplast proteins under oxidative stress conditions (Ito et al. 2003; Zaffagnini et al. 2007, 2012a, b). The glutathionylation reaction is generally supported by ROS such as H_2O_2 under stress conditions (Zaffagnini et al. 2012b). In the absence of a glutathionylation reaction, the thiol group of cysteine could be oxidized to irreversible forms, i.e., sulfinates and sulfonates (Poole et al. 2004). In this way, the reaction of GSH with thiol groups of cysteine (glutathionylation) protects proteins from possible damage by ROS on redox signaling, although it has yet to be completely elucidated and is currently under extensive investigation (Zaffagnini et al. 2012a).

A number of redoxactive enzymes are known to intervene in the glutathionylation process. Examples, on which we elaborate below, are the peroxiredoxins (PRDXs) (Dietz 2003; Zaffagnini et al. 2012a), glutaredoxins (GRXs) (Xing et al. 2006; Meyers et al. 2008), thioredoxins (TRXs) (Buchanan and Balmer 2005; Zaffagnini et al. 2012a), and protein disulfide isomerases (Alergand et al. 2006). These redoxactive enzymes, together with a various redox-active target proteins defend proteins from irreversible oxidation especially under oxidative stress conditions (Ströher and Dietz 2006; Meyers et al. 2008; Zaffagnini et al. 2012a).

Peroxiredoxin (PRDXs) comprises a family of thiol-based peroxidases found in organisms ranging from bacteria to mammals (Abbas et al. 2008; Bhatt and Tripathi 2011; Anjum et al. 2012; Djuika et al. 2013). Though the roles of PRDXs have not yet been completely elucidated, their role in heavy-metal-induced ROS detoxification is evident (Matamoros et al. 2010; Abbas et al. 2013). The proteomic analysis of maize roots (Requejo up-regulation of PRDXs under heavy metal stress. These enzymes usually catalyze the reduction of H_2O_2 and other hydroperoxides (ROOH) with help from reduced thioredoxins, to yield thioredoxin disulfide, water, and the corresponding alcohol (Dietz 2011; Deponte 2013; Djuika et al. 2013; Randall et al. 2013). Bhatt and Tripathi (2011) described the reaction mechanism of PRDXs-induced decomposition of $O_2^{\cdot-}$ to H_2O. They summarized the entire process in three steps: peroxidation, redox dehydration and reduction as reported by Aran et al. (2009). The reaction starts as a nucleophilic attack of the protein thiol on the peroxide, resulting in the release of an alcohol and concomitant oxidation to a sulfenic acid (RSOH), which starts the catalytic cycle (Ellis and Poole 1997). The thiol group of Cys attacks RSOH, resulting in the release of H_2O and formation of a disulfide bridge. The catalytic cycle is stopped by a complementary reduction system, which results in catalytically active PRDXs (Aran et al. 2009; Bhatt and Tripathi 2011). Peroxiredoxin with CAT and other peroxidases are reported to take part in signal transduction by controlling the intracellular H_2O_2 concentration (Randall et al. 2013; Poynton and Hampton 2013). In plants, PRDXs have four subgroups (1-Cys PRDX, 2-Cys PRDX, PRDX II and PRDX Q) that are based on the number and position of the conserved cysteine residues, genome-wide analysis of plants and their subunit composition (Rouhier et al. 2001; Rouhier and Jacquot 2002; Poynton and Hampton 2013).

Thioredoxin (TRXs) is a family of antioxidant redox proteins (12.4 kDa) that facilitate the reduction of other proteins through the exchange of thiol/disulfide (Lemaire et al. 2003). For example, thioredoxins act as hydrogen donors for thioredoxin peroxidases or peroxiredoxin, which are involved in the removal of H_2O_2 (Verdoucq et al. 1999; Behm and Jacquot 2000). The reaction mechanism involves the reduction of the oxidized disulfide form of thioredoxin by NADPH and thioredoxin reductase (TRR). Depending on the primary sequence and sub-cellular localization, plants have six subgroups/types (TRXs f, m, x, y, h, and o). These subgroups have different sub-cellular compartmentalization and function. Thioredoxin-x, -y, -z, and NTRc are reported to act as electron donors to various antioxidant enzymes such as the glutathione peroxidises, methionine sulfoxide reductases and peroxiredoxins (Tarrago et al. 2009; Chibani et al. 2010).

However, it is not always evident that ROS detoxification by antioxidant enzymes requires electrons from the glutaredoxin or thioredoxin systems (Culotta et al. 2006; Benabdellah et al. 2009). It is reported that in GSH deficient cells, TRXs are overproduced to compensate for GSH shortage (Pócsi et al. 2004). Examination of the redox state of TRXs and GRXs in mutant plants showed that TRXs are independent of the GSH/GRX system (Trotter and Grant 2003). Still the interaction of TRXs,

GRXs and GSH in redox-dependent regulation, based on disulphide/dithiol exchange reactions under stress conditions (overproduction of ROS), is not well established in plants.

Glutaredoxins (GRXs) are oxidoreductases that catalyze the reversible reduction of disulfide bonds and participate in antioxidant defence by reducing various enzymes such as peroxiredoxins, dehydroascorbate, and methionine sulfoxide reductase (Buchanan and Balmer 2005; Li and Zachgo 2009). Glutaredoxins are oxidized by substrates, and reduced non-enzymatically by GSH. In the dithiol mechanism, electrons are transferred from NADPH to GR, then to GSH, and from there to GRXs. Finally, GRXs reduce target proteins by dithiol-disulfide exchange reactions in a manner similar to TRXs. The plant glutaredoxin family contains more than 30 members that are localized in different cell compartments (Couturier et al. 2009; Zaffagnini et al. 2012b). Almost thirty different GRXs isoforms have been identified in *A. thaliana*. They are subgrouped in six classes based on their redox-active center (Xing et al. 2006). Each class contains a variant of the active site motif and peculiar functional properties (Rouhier et al. 2006). GPXs appears to be involved in detoxifying H_2O_2 (Foyer and Noctor 2005, 2009) as well as lipid and phospholipid hydroperoxides (Avery and Avery 2001). GRXs also participate to reduce the oxidized cysteines, providing evidence of GRXs role in oxidative stress signaling (Michelet et al. 2006).

6.4 Nitrogen Metabolism

Nitrogen metabolism plays an important role in plant responses to heavy metal toxicity (Lea and Azevedo 2007; Andrade et al. 2010). Various nitrogenous metabolites, such as polyamines, amino acids and amino acid-derived molecules can bind to and scavenge heavy-metal-induced ROS (Kovac et al. 2009; Radić et al. 2010). When plants are exposed to high heavy metals levels, it is reported that some plant amino acids (e.g., proline or histidine), scavenge ROS (Sharma and Dietz 2006; Fariduddin et al. 2009).

Huang and Wang (2010) suggested that free prolines help protect certain plant enzymes, osmoregulation and help to stabile the sub-cellular components and structures. Proline has been reported to accumulate in plants under heavy metal stress conditions, an indication that its increased presence provides a protective or a regulatory role (Sharma and Dietz 2006). Metal-tolerant plants contain higher constitutive proline levels, even in the absence of excess metal ions, than do non-tolerant plants (Sharma and Dietz 2006; Huang and Wang 2010). Increased levels of proline correlate with enhanced metal tolerance in plants, and as a result, some researchers believe it to act as an antioxidant in metal-stressed cells (Gupta and Sarin 2009; Huang and Wang 2010). One of the proposed roles of proline is to reduce free radical levels that are generated from toxicity events. In this regard, proline acts in a manner that is similar to GSH, ascorbic acid or tocopherol. Heavy metals interfere with N metabolism to cause toxicity that alters the composition of amino acid in plants (Callahan et al. 2007).

6.5 Antioxidant Enzymes

One of the most efficient mechanisms that plants use to protect themselves is to detoxify any free radicals that are present. Such detoxification prevents cell injury and tissue dysfunction and is accomplished in plant cells via activation of antioxidants enzymes such as SOD, CAT, POD, APX, GR, DHAR and MDHAR (Table 2, Fig. 4) (Lomonte et al. 2010; Mou et al. 2011; Vanhoudt et al. 2011; Lyubenova and Schröder 2011; Cestone et al. 2012; Opdenakker et al. 2012; Shahid et al. 2013d). Previous results have shown that high levels of antioxidant enzymes can increase stress tolerance to heavy-metal-induced stress conditions. Many researchers have also reported that antioxidant enzymes are activated in different plant species to scavenge the ROS that are produced by heavy metal toxicity (Gonnelli et al. 2001; Kim et al. 2010; Kafel et al. 2010; Martínez Domínguez et al. 2010; He et al. 2011).

Plant species display different levels of tolerance to heavy metal exposure (Shahid et al. 2012d), and the enzymes in these plants display varying behavior when under heavy metal stress. Most of these antioxidative enzymes are electron donors and react with free radicals to form innocuous end products, such as water. The process involves the binding of these ROS to active enzyme sites, and then conversion to non-toxic and inactive products. Among these enzymes, SOD is a key one for defending plants against ROS. The catalytic properties of SOD were first detected by McCord and Fridovich (1969). SOD is responsible for dismutation of the two superoxide radicals to H_2O_2 and O_2. In this way, SOD maintains $O_2^{\cdot-}$ at a steady state level (Gao et al. 2010; Deng et al. 2010; Andrade et al. 2010; Cestone et al. 2012). An increase in SOD activity could be either direct through the action of heavy metal ions on SOD, or indirect through an increase in $O_2^{\cdot-}$ levels (Chongpraditnun et al. 1992; Shahid et al. 2013d). When SOD appears, it generally does so in response to the production of heavy-metal-induced H_2O_2, which can form lipid peroxides by direct or indirect action by lipoxygenase- mediated lipid peroxidation (Deng et al. 2010). An increase in SOD activity may result from enhanced formation of $O_2^{\cdot-}$ or from de novo synthesis of enzyme proteins (Verma and Dubey 2003; Yılmaz and Parlak 2011). Catalase is generally present in mitochondria and peroxisomes, where it decomposes H_2O_2 to H_2O and O_2 (Hermes-Lima 2005; Tang et al. 2010; Shahid et al. 2013d). Another enzyme class responsible for degrading H_2O_2 are the PODs, which are capable of reducing H_2O_2 to H_2O. Guaiacol peroxidase is present in vacuoles, the cell wall, cytosol and extracellular spaces. POD is considered to be a marker of heavy metal toxicity, having broad specificity for phenolic substrates and higher affinity for H_2O_2 than CAT (Radwan et al. 2010). Guaiacol peroxidase consumes H_2O_2 to generate phenoxy compounds that are polymerized to produce cell wall components such as lignin (Mishra et al. 2006; Pourrut et al. 2011b).

Enzymes of ascorbate–glutathione cycle, APX and GR, are located mainly in chloroplasts, other cellular organelles and the cytoplasm, where they are involved in controlling the cellular redox status, especially under heavy metals stress conditions (Singh et al. 2010). Ascorbic acid is a primary and secondary antioxidant. APX utilizes ascorbate to reduce H_2O_2 to H_2O and O_2 (Mittler 2002; Triantaphylidès and Havaux 2009). During this process, ascorbate is oxidized to monodehydroascorbate.

Table 2 The antioxidant enzyme systems different plants use to defend themselves against heavy-metal-induced ROS

Heavy metals	Enzymes	Plant species	References
Ag	SOD, CAT	*Potamogeton crispus*	Xu et al. (2010b)
Al	SOD, CAT, APX, GPOX	*Hordeum vulgare*	Achary et al. (2012)
	SOD, POD	*Hordeum vulgare*	Guo et al. (2007)
As	SOD, GR, SDH	*Aspergillus niger*	Mukherjee et al. (2010)
	SOD, POD, APX, CAT	*Zea mays, Vicia faba*	Duquesnoy et al. (2010)
	APX, MDHAR, DHAR, SOD, GST	*Typha latifolia*	Lyubenova and Schröder (2011)
Cd	SOD, POD, CAT	*Carassius auratus*	Chen et al. (2012)
	SOD, APX, GR	*Gracilaria dura*	Kumar et al. (2012)
	APX, MDHAR, DHAR, GR, GST	*Helianthus annuus*	Nehnevajova et al. (2012)
	SOD, CAT, APX, GR	*Solanum lycopersicum*	Cherif et al. (2011)
	APX, MDHAR, DHAR, SOD, GST	*Typha latifolia*	Lyubenova and Schröder (2011)
	SOD, APX, CAT, GR	*Brassica juncea*	Ahmad et al. (2011b)
	SOD, POD, CAT	*Medicago sativa*	Xu et al. (2010a)
	POD, CAT	*Amaranthus hybridus*	Zhang et al. (2010)
	GSH, GST	*Brassica juncea*	Szőllősi et al. (2009)
	SOD, POD	*Hordeum vulgare*	Guo et al. (2007)
Cr	GPX, APX, CAT, GR	*Zea mays*	Mallick et al. (2010)
	APX, SOD, POD	*Lycopersicum esculatum*	Nayek et al. (2010)
Cu	APX, MDHAR, DHAR, GR, GST	*Helianthus annuus*	Nehnevajova et al. (2012)
	SOD, CAT, APX	*Pisum sativum*	Turchi et al. (2012)
	SOD, APX, GR	*Sesbania drummondii*	Israr et al. (2011)
	GPX, CAT	*Phaseolusvulgaris*	Bouazizi et al. (2010)
	SOD, POD, CAT	*Vetiveria zizanioides*	Xu et al. (2009)
	SOD, POD, APX, CAT	*Withania somnifera*	Khatun et al. (2008)
	SOD, GPX, CAT	*Datura stramonium* *Malva sylvestris* *Chenopodium ambrosioides*	Boojar and Goodarzi (2007)
	SOD, POD	*Hordeum vulgare*	Guo et al. (2007)
Ni	SOD, CAT, APX, GPOX, GR	*Brassica juncea*	Kanwar et al. (2012)
	SOD, APX, GR	*Sesbania drummondii*	Israr et al. (2011)
Pb	SOD	*Spinacia oleracea*	Wang et al. (2010)
	APX, MDHAR, DHAR, SOD, GST	*Typha latifolia*	Lyubenova and Schröder (2011)
	SOD, APX	*Sedum alfredii*	Gupta et al. (2010)
	SOD, GPX, APX, CAT, GR	*Najas indica*	Sing et al. (2010)
	SOD, APX, GR	*Sesbania drummondii*	Israr et al. (2011)
	APX, SOD, POD	*Lycopersicum esculatum*	Nayek et al. (2010)
	SOD, CAT, AsA	*Zea mays*	Gupta et al. (2009)
	APX, GR, GST	*Lathyrus sativus*	Brunet et al. (2009)
	CAT, APX	*Wolffia arrhiza*	Piotrowska et al. (2009)
	APX, SOD, POD	*Lycopersicum esculatum*	Nayek et al. (2010)
Zn	APX, MDHAR, DHAR, GR, GST	*Helianthus annuus*	Nehnevajova et al. (2012)
	SOD, CAT, APX	*Pisum sativum*	Turchi et al. (2012)
	SOD, CAT, APX, GR	*Solanum lycopersicum*	Cherif et al. (2011)
	SOD, APX, GR	*Sesbania drummondii*	Israr et al. (2011)
	SOD, POD, CAT	*Vetiveria zizanioides*	Xu et al. (2009)

SOD superoxide dismutase, *APX* ascorbate peroxidise, *GPX* guaiacol peroxidise, *CAT* catalase, *GR* glutathione reductase, *AsA* ascorbic acid, *GSH* glutathione, *GST* glutathione S-transferase, *POD* peroxidase, *DHAR* dehydroascorbate; reductase, *MDHAR* monodehydroascorbate reductase, *ACOX* acyl co-A oxidase, *SDH* succinatedehydrogenase

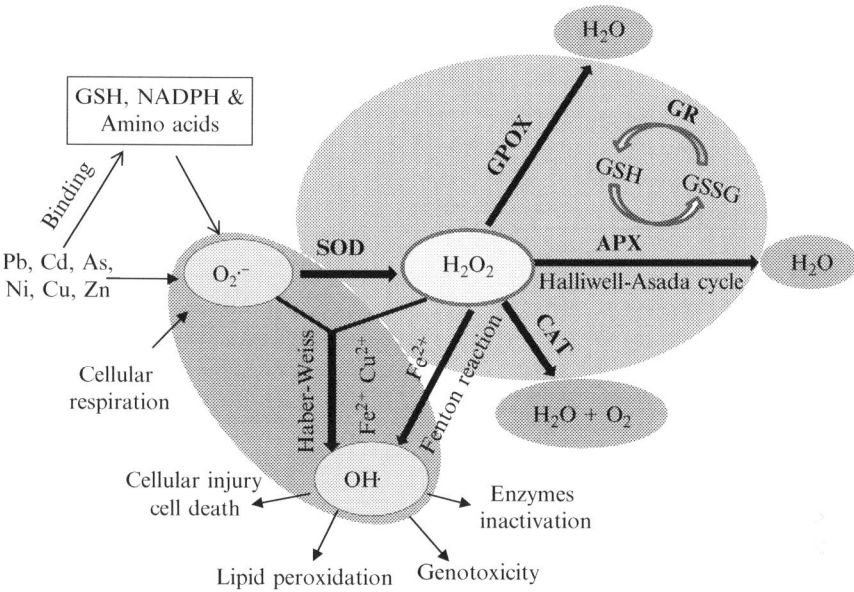

Fig. 4 Schematic representation of heavy-metal-induced oxidative stress. Under normal conditions (*highlighted grey*), $O_2^{\cdot-}$ is produced by cellular respiration. This $O_2^{\cdot-}$ is converted to H_2O_2 by SOD. The H_2O_2 produced is converted to H_2O and O_2 by the combined action of APX, GPOX, CAT and GR. In the presence of heavy metals, the $O_2^{\cdot-}$ and H_2O_2 production is increased. The increased ROS is incompletely converted to H_2O by the antioxidants. As a result, highly toxic HO˙ is produced by the Haber–Weiss or Fenton reactions. This HO˙ is the most toxic ROS and is believed to initiate lipid peroxidation, cell death, enzyme inactivation and genotoxicity

The monodehydroascorbate formed can be directly reduced back to ascorbate by monodehydroascorbate reductase (MDHAR), or may first be converted to dehydroascorbate, and then reduced by dehydroascorbate reductase (DHAR). In the process, GSH acts as reductant, which is oxidized to GSSG (oxidized glutathione). When GR activity is induced, the GSH/GSSG ratio remains high, and thus allows GSH to participate in PC synthesis and ROS detoxification (Noctor et al. 1998).

Several previous authors have reported heavy-metal-induced increases in antioxidant enzymes (Table 2). Ali et al. (2011) observed activation of SOD, POD, APX, GR and CAT under Al or Cr stress in *Hordeum vulgare*. Israr et al. (2011) reported a significant increase in enzymatic (SOD, APX, GR) antioxidant levels in *Sesbania drummondii* seedlings, when the seedlings were exposed to Cu, Ni and Zn alone and in combination. Lomonte et al. (2010) reported increased CAT and SOD activity, in response to applying Hg to *Atriplex codonocarpa* for 4 weeks under hydroponic conditions. Radić et al. (2010) also reported increased SOD and POD activity, when *Lemna minor* plants were exposed to Al and Zn. Yadav (2010) observed that the antioxidants CAT, APX and glutathione S-transferase (GST) increased as the Cr concentration increased in *Jatropha curcas*. Shahid (2010) reported a Pb-induced increase in APX, SOD, GPX and GR levels in *Vicia faba* roots and leaves, as did (Choudhary et al. 2010) in *Raphanus sativus* by Cu.

Increased activity of POD and CAT in *Amaranthus hybridus*, in reponse to Cd toxicity, was also observed by Zhang et al (2010). Singh et al. (2010) reported that the bioaccumulation of Pb by *Najas indica* activated several antioxidant enzymes (e.g., SOD, APX, GPX, CAT and GR). They also reported significantly increased cysteine synthase and glutathione-S-transferase activity. Similar results have been reported for *Phaseolus aureus* and *Vicia sativa* (Zhang et al. 2009). Recently, Shahid (2010) reported the results of a time course experiment (1, 4, 8; 12 and 24 h), in which the Pb-induced activation of antioxidant enzymes (APX, GPOX, SOD and GR), lipid peroxidation and ROS production occurred, after the Pb concentration reached significant levels in roots (after 1 h) and leaves (after 8 h). This suggests that Pb-induced lipid peroxidation, activation of enzymes and production of H_2O_2 are very rapid phenomena. Moreover, the oxidative bursts in roots and leaves coincide with periods of high Pb entrance rates to these tissues (1 and 12 h) (Pourrut et al. 2008).

7 Conclusions and Perspectives

In this review, we have highlighted key results from the previous and particular the recent published literature that addresses heavy-metal-induced physiological changes that occur in plants. Based on the literature cited in this review, we have drawn the following conclusions:

1. The generation of ROS is an inevitable feature of higher plants and other aerobic organisms. These ROS are constantly generated as side-products of certain metabolic pathways, and act to control various essential plant processes. Heavy metal exposure to plants disturbs the delicate balance between ROS production and elimination, leading to an enhanced steady-state ROS level that is called "**oxidative stress**". A common feature of oxidative stress is damage to proteins, DNA, and lipids. Consequently, it is suggested that metal-induced oxidative stress in cells may partially be responsible for the toxic effects produced by heavy metals.
2. The plant kingdom has evolved a very efficient enzymatic and nonenzymatic defense system that allows ROS-scavenging to protect plant cells from oxidative damage. Retention of heavy metals in the cell wall is the first barrier against heavy metal stress. Heavy metal chelation by PCs, MTs, GSH and amino acids, and subsequent sequestration in vacuoles is another detoxification mechanism in plants. Biochemical tolerance to heavy metals is linked to activation of antioxidant enzymes. These heavy metal tolerance mechanisms may be activated separately or simultaneously, depending on the type and species of metal and plant.
3. ROS-induced toxicity to different plant molecules and the various responses of plants to over production of ROS are often used as bioindicators in risk and environmental quality assessment studies. Such biomarkers are appropriate for use in ecotoxicological studies. To further develop and improve these bioindicators, a better understanding of the processes and mechanisms involved in ROS production, their toxicity and defense mechanisms in the presence of pollutants, such as

heavy metals, are needed. Moreover, all bioindicators are not equally sensitive to different pollutants under different environmental conditions. Therefore, the mechanisms behind ROS production, toxicity and detoxification should be compared to optimize the most sensitive and efficient assays, with respect to environmental conditions like applied metal form and concentration, physico-chemical parameters of medium and metal and plant type.

8 Summary

As a result of the industrial revolution, anthropogenic activities have enhanced the redistribution of many toxic heavy metals from the earth's crust to different environmental compartments. Environmental pollution by toxic heavy metals is increasing worldwide, and poses a rising threat to both the environment and to human health. Plants are exposed to heavy metals from various sources: mining and refining of ores, fertilizer and pesticide applications, battery chemicals, disposal of solid wastes (including sewage sludge), irrigation with wastewater, vehicular exhaust emissions and adjacent industrial activity.

Heavy metals induce various morphological, physiological, and biochemical dysfunctions in plants, either directly or indirectly, and cause various damaging effects. The most frequently documented and earliest consequence of heavy metal toxicity in plants cells is the overproduction of ROS. Unlike redox-active metals such as iron and copper, heavy metals (e.g. Pb, Cd, Ni, Al, Mn and Zn) cannot generate ROS directly by participating in biological redox reactions such as Haber–Weiss/Fenton reactions. However, these metals induce ROS generation via different indirect mechanisms, such as stimulating the activity of NADPH oxidases, displacing essential cations from specific binding sites of enzymes and inhibiting enzymatic activities from their affinity for –SH groups on the enzyme.

Under normal conditions, ROS play several essential roles in regulating the expression of different genes. Reactive oxygen species control numerous processes like the cell cycle, plant growth, abiotic stress responses, systemic signalling, programmed cell death, pathogen defence and development. Enhanced generation of these species from heavy metal toxicity deteriorates the intrinsic antioxidant defense system of cells, and causes oxidative stress. Cells with oxidative stress display various chemical, biological and physiological toxic symptoms as a result of the interaction between ROS and biomolecules. Heavy-metal-induced ROS cause lipid peroxidation, membrane dismantling and damage to DNA, protein and carbohydrates. Plants have very well-organized defense systems, consisting of enzymatic and non-enzymatic antioxidation processes. The primary defense mechanism for heavy metal detoxification is the reduced absorption of these metals into plants or their sequestration in root cells. Secondary heavy metal tolerance mechanisms include activation of antioxidant enzymes and the binding of heavy metals by phytochelatins, glutathione and amino acids. These defense systems work in combination to manage the cascades of oxidative stress and to defend plant cells from the toxic effects of ROS.

In this review, we summarized the biochemical processes involved in the overproduction of ROS as an aftermath to heavy metal exposure. We also described the ROS scavenging process that is associated with the antioxidant defense machinery. Despite considerable progress in understanding the biochemistry of ROS overproduction and scavenging, we still lack in-depth studies on the parameters associated with heavy metal exclusion and tolerance capacity of plants. For example, data about the role of glutathione–glutaredoxin–thioredoxin system in ROS detoxification in plant cells are scarce. Moreover, how ROS mediate glutathionylation (redox signalling) is still not completely understood. Similarly, induction of glutathione and phytochelatins under oxidative stress is very well reported, but it is still unexplained that some studied compounds are not involved in the detoxification mechanisms. Moreover, although the role of metal transporters and gene expression is well established for a few metals and plants, much more research is needed. Eventually, when results for more metals and plants are available, the mechanism of the biochemical and genetic basis of heavy metal detoxification in plants will be better understood. Moreover, by using recently developed genetic and biotechnological tools it may be possible to produce plants that have traits desirable for imparting heavy metal tolerance.

Acknowledgement The authors thank the Higher Education Commission of Pakistan (http://www.hec.gov.pk) and the French Society for Export of Educative Resources (SFERE, http://www.sfere.fr/) for the scholarship granted to M. Shahid.

References

Abbas K, Breton J, Drapier J-C (2008) The interplay between nitric oxide and peroxiredoxins. Immunobiology 213:815–822

Abbas K, Riquier S, Drapier J-C (2013) Peroxiredoxins and sulfiredoxin at the crossroads of the NO and H_2O_2 signaling pathways. Methods Enzymol 527:113–128

Achard P, Renou J-P, Berthomé R, Harberd NP, Genschik P (2008) Plant DELLAs restrain growth and promote survival of adversity by reducing the levels of reactive oxygen species. Curr Biol 18:656–660

Achary MMV, Patnaik AR, Panda BB (2012) Oxidative biomarkers in leaf tissue of barley seedlings in response to aluminum stress. Ecotoxicol Environ Saf 75:16–26

Ahmad MSA, Ashraf M (2011) Essential roles and hazardous effects of nickel in plants. Rev Environ Contam Toxicol 214:125–167

Ahmad MSA, Ashraf M, Hussain M (2011a) Phytotoxic effects of nickel on yield and concentration of macro- and micro-nutrients in sunflower (*Helianthus annuus* L.) achenes. J Hazard Mater 185:1295–1303

Ahmad P, Nabi G, Ashraf M (2011b) Cadmium-induced oxidative damage in mustard [*Brassica juncea* (L.) Czern. & Coss.] plants can be alleviated by salicylic acid. S Afr J Bot 77:36–44

Ahsan N, Lee D-G, Lee S-H, Kang KY, Lee JJ, Kim PJ, Yoon H-S, Kim J-S, Lee B-H (2007) Excess copper induced physiological and proteomic changes in germinating rice seeds. Chemosphere 67:1182–1193

Aina R, Sgorbati S, Santagostino A, Labra M, Ghiani A, Citterio S (2004) Specific hypomethylation of DNA is induced by heavy metals in white clover and industrial hemp. Physiol Plant 121:472–480

Alergand T, Peled-Zehavi H, Katz Y, Danon A (2006) The chloroplast protein disulfide isomerase RB60 reacts with a regulatory disulfide of the RNA-binding protein RB47. Plant Cell Physiol 47:540–548

Ali MF, Ahmed S, Qureshi MS (2011) Catalytic coprocessing of coal and petroleum residues with waste plastics to produce transportation fuels. Fuel Process Technol 92:1109–1120

Alméras E, Stolz S, Vollenweider S, Reymond P, Mène-Saffrané L, Farmer EE (2003) Reactive electrophile species activate defense gene expression in *Arabidopsis*. Plant J 34:205–216

Álvarez R, Hoyo AD, García-Breijo F, Reig-Arminana J, del Campo EM, Guéra A, Barreno E, Casano LM (2012) Different strategies to achieve Pb-tolerance by the two Trebouxia algae coexisting in the lichen *Ramalina farinacea*. J Plant Physiol. doi:org/10.1016/j.jplph.2012.07.005

An J, Jeong S, Moon HS, Jho EH, Nam K (2012) Prediction of Cd and Pb toxicity to *Vibrio fischeri* using biotic ligand-based models in soil. J Hazard Mater 203–204:69–76

Anastassopoulou J (2003) Metal–DNA interactions. J Mol Struct 651–653:19–26

Andra SS, Datta R, Sarkar D, Makris KC, Mullens CP, Sahi SV, Bach SBH (2009a) Synthesis of phytochelatins in vetiver grass upon lead exposure in the presence of phosphorus. Plant Soil 326:171–185

Andra SS, Datta R, Sarkar D, Saminathan SKM, Mullens CP, Bach SBH (2009b) Analysis of phytochelatin complexes in the lead tolerant vetiver grass [*Vetiveria zizanioides* (L.)] using liquid chromatography and mass spectrometry. Environ Pollut 157:2173–2183

Andrade LR, Farina M, Amado Filho GM (2004) Effects of copper on *Enteromorpha flexuosa* (Chlorophyta) in vitro. Ecotoxicol Environ Saf 58:117–125

Andrade SAL, Gratao PL, Azevedo RA, Silveira APD, Schiavinato MA, Mazzafera P (2010) Biochemical and physiological changes in jack bean under mycorrhizal symbiosis growing in soil with increasing Cu concentrations. Environ Exp Bot 68:198–207

Anjum NA, Ahmad I, Mohmood I, Pacheco M, Duarte AC, Pereira E, Umar S, Ahmad A, Khan NA, Iqbal M, Prasad MNV (2012) Modulation of glutathione and its related enzymes in plants' responses to toxic metals and metalloids—A review. Environ Exp Bot 75:307–324

Antolín MC, Muro I, Sánchez-Díaz M (2010) Sewage sludge application can induce changes in antioxidant status of nodulated alfalfa plants. Ecotoxicol Environ Saf 73:436–442

Aran M, Ferrero DS, Pagano E, Wolosiuk RA (2009) Typical 2-Cys peroxiredoxins–modulation by covalent transformations and noncovalent interactions. FEBS J 276:2478–2493

Arasimowicz-Jelonek M, Floryszak-Wieczorek J, Gwóźdź EA (2011) The message of nitric oxide in cadmium challenged plants. Plant Sci 181:612–620

Aravind P, Prasad MNV, Malec P, Waloszek A, Strzałka K (2009) Zinc protects *Ceratophyllum demersum* L. (free-floating hydrophyte) against reactive oxygen species induced by cadmium. J Trace Elem Med Biol 23:50–60

Arduini I, Godbold DL, Onnis A (1995) Influence of copper on root growth and morphology of *Pinus pinea* L. and *Pinus pinaster* seedlings. Tree Physiol 15:411–415

Arias JA, Peralta-Videa JR, Ellzey JT, Ren M, Viveros MN, Gardea-Torresdey JL (2010) Effects of *Glomus deserticola* inoculation on Prosopis: Enhancing chromium and lead uptake and translocation as confirmed by X-ray mapping, ICP-OES and TEM techniques. Environ Exp Bot 68:139–148

Arshad M, Silvestre J, Pinelli E, Kallerhoff J, Kaemmerer M, Tarigo A, Shahid M, Guiresse M, Pradere P, Dumat C (2008) A field study of lead phytoextraction by various scented Pelargonium cultivars. Chemosphere 71:2187–2192

Averill-Bates DA, Przybytkowski E (1994) The role of glucose in cellular defences against cytotoxicity of hydrogen peroxide in Chinese hamster ovary cells. Arch Biochem Biophys 312:52–58

Avery AM, Avery SV (2001) *Saccharomyces cerevisiae* expresses three phospholipid hydroperoxide glutathione peroxidases. J Biol Chem 276:33730–33735

Bai C, Reilly CC, Wood BW (2006) Nickel deficiency disrupts metabolism of ureids, amino acids and organic acids of young pecan foliage. Plant Physiol 140:433–443

Bai J, Xiao R, Cui B, Zhang K, Wang Q, Liu X, Gao H, Huang L (2011) Assessment of heavy metal pollution in wetland soils from the young and old reclaimed regions in the Pearl River Estuary, South China. Environ Pollut 159:817–824

Bailly C, El-Maarouf-Bouteau H, Corbineau F (2008) From intracellular signaling networks to cell death: the dual role of reactive oxygen species in seed physiology. C R Biol 331:806–814

Bal W, Kasprzak KS (2002) Induction of oxidative DNA damage by carcinogenic metals. Toxicol Lett 127:55–62

Bannister JV, Halliwell B, O'Neill P (1985) Free radicals in biological medicine, vol 3. Harwood Academic, London, pp 1–266

Barbosa JS, Cabral TM, Ferreira DN, Agnez-Lima LF, De Medeiros SRB (2010) Genotoxicity assessment in aquatic environment impacted by the presence of heavy metals. Ecotoxicol Environ Saf 73:320–325

Barranco-Medina S, Krell T, Finkemeier I, Sevilla F, Lázaro J-J, Dietz K-J (2007) Biochemical and molecular characterization of the mitochondrial peroxiredoxin PsPrxII F from *Pisum sativum*. Plant Physiol Biochem 45:729–739

Bartoli CG, Casalongué CA, Simontacchi M, Marquez-Garcia B, Foyer CH (2012) Interactions between hormone and redox signalling pathways in the control of growth and cross tolerance to stress. Environ Exp Bot. doi:10.1016/j.envexpbot.2012.05.003

Becana M, Dalton DA, Moran JF, Iturbe-Ormaetxe I, Matamoros MA, Rubio CM (2000) Reactive oxygen species and antioxidants in legume nodules. Physiol Plant 109:372–381

Behm M, Jacquot J-P (2000) Isolation and characterization of thioredoxin h from poplar xylem. Plant Physiol Biochem 38:363–369

Beltagi MS (2005) Phytotoxicity of lead (Pb) to SDS-PAGE protein profile in root nodules of faba bean (*Vicia faba* L.) plants. Pak J Biol Sci 8:687–690

Benabdellah K, Merlos M-A, Azcón-Aguilar C, Ferrol N (2009) GintGRX1, the first characterized glomeromycotan glutaredoxin, is a multifunctional enzyme that responds to oxidative stress. Fungal Genet Biol 46:94–103

Benekos K, Kissoudis C, Nianiou-Obeidat I, Labrou N, Madesis P, Kalamaki M, Makris A, Tsaftaris A (2010) Overexpression of a specific soybean GmGSTU4 isoenzyme improves diphenyl ether and chloroacetanilide herbicide tolerance of transgenic tobacco plants. J Biotechnol 150:195–201

Bhatt I, Tripathi BN (2011) Plant peroxiredoxins: catalytic mechanisms, functional significance and future perspectives. Biotechnol Adv 29:850–859

Bhattacharjee S (2005) Reactive oxygen species and oxidative burst: roles in stress, senescence and signal transduction in plants. Curr Sci 89:1113–1121

Bhattacharjee S (2012) The language of reactive oxygen species signaling in plants. J Bot 2012:1–22

Blokhina O, Fagerstedt KV (2010) Oxidative metabolism, ROS and NO under oxygen deprivation. Plant Physiol Biochem 48:359–373

Boojar M, Goodarzi F (2007) The copper tolerance strategies and the role of antioxidative enzymes in three plant species grown on copper mine. Chemosphere 67:2138–2147

Borisova MM, Kozuleva MA, Rudenko NN, Naydov IA, Klenina IB, Ivanov BN (2012) Photosynthetic electron flow to oxygen and diffusion of hydrogen peroxide through the chloroplast envelope via aquaporins. Biochim Biophys Acta 1817:1314–1321

Bouazizi H, Jouili H, Geitmann A, El Ferjani E (2010) Copper toxicity in expanding leaves of *Phaseolus vulgaris* L.: antioxidant enzyme response and nutrient element uptake. Ecotoxicol Environ Saf 73:1304–1308

Buchanan BB, Balmer Y (2005) Redox regulation: a broadening horizon. Annu Rev Plant Biol 56:187–220

Buchanan BB, Gruissem W, Jones RL (2000) Biochemistry and molecular biology of plants. American Society of Plant Physiologist, Rockville, MD

Callahan D, Kolev S, O'Hair R, Salt D, Baker A (2007) Relationships of nicotianamine and other amino acids with nickel, zinc and iron in *Thlaspi hyperaccumulators*. New Phytol 176:836–848

Carrasco-Gil S, Estebaranz-Yubero M, Medel-Cuesta D, Millán R, Hernández LE (2012) Influence of nitrate fertilization on Hg uptake and oxidative stress parameters in alfalfa plants cultivated in a Hg-polluted soil. Environ Exp Bot 75:16–24

Cenkci S, Yildiz M, Ciğerci IH, Konuk M, Bozdağ A (2009) Toxic chemicals-induced genotoxicity detected by random amplified polymorphic DNA (RAPD) in bean (*Phaseolus vulgaris* L.) seedlings. Chemosphere 76:900–906

Cestone B, Cuypers A, Vangronsveld J, Sgherri C, Navari-Izzo F (2012) The influence of EDDS on the metabolic and transcriptional responses induced by copper in hydroponically grown *Brassica carinata* seedlings. Plant Physiol Biochem 55:43–51

Chaffai R, Koyama H (2011) Heavy metal tolerance in *Arabidopsis thaliana*. In: Kader J-C, Delseny M (eds) Advances in botanical research. Academic, London, pp 1–49, Chapter 1

Chatterjee C, Dube BK, Sinha P, Srivastava P (2004) Detrimental effects of lead phytotoxicity on growth, yield, and metabolism of rice. Commun Soil Sci Plant Anal 35:255–265

Chen J, Zhu C, Li L-P, Z-yang S, X-bo P (2007) Effects of exogenous salicylic acid on growth and H_2O_2-metabolizing enzymes in rice seedlings under lead stress. J Environ Sci (China) 19: 44–49

Chen F, Gao J, Zhou Q (2012) Toxicity assessment of simulated urban runoff containing polycyclic musks and cadmium in *Carassius auratus* using oxidative stress biomarkers. Environ Pollut 162:91–97

Cherif J, Mediouni C, Ammar WB, Jemal F (2011) Interactions of zinc and cadmium toxicity in their effects on growth and in antioxidative systems in tomato plants (*Solarium lycopersicum*). J Environ Sci 23:837–844

Chibani K, Couturier J, Selles B, Jacquot J-P, Rouhier N (2010) The chloroplastic thiol reducing systems: dual functions in the regulation of carbohydrate metabolism and regeneration of antioxidant enzymes, emphasis on the poplar redoxin equipment. Photosynth Res 104:75–99

Chongpraditnun P, Mori S, Chino M (1992) Excess copper induces a cytosolic Cu, Zn-Superoxide dismutase in soybean root. Plant Cell Physiol 33:239–244

Choudhary SP, Bhardwaj R, Gupta BD, Dutt P, Gupta RK, Biondi S, Kanwar M (2010) Epibrassinolide induces changes in indole-3-acetic acid, abscisic acid and polyamine concentrations and enhances antioxidant potential of radish seedlings under copper stress. Physiol Plant 140:280–296

Circu ML, Aw TY (2010) Reactive oxygen species, cellular redox systems, and apoptosis. Free Radic Biol Med 48:749–762

Clemens S (2006) Toxic metal accumulation, responses to exposure and mechanisms of tolerance in plants. Biochimie 88:1707–1719

Collin-Hansen C, Andersen RA, Steinnes E (2005) Damage to DNA and lipids in *Boletus edulis* exposed to heavy metals. Mycol Res 109:1386–1396

Cooke MS, Evans MD, Dizdaroglu M, Lunec J (2003) Oxidative DNA damage: mechanisms, mutation, and disease. FASEB J 17:1195–1214

Corpas FJ, Leterrier M, Valderrama R, Airaki M, Chaki M, Palma JM, Barroso JB (2011) Nitric oxide imbalance provokes a nitrosative response in plants under abiotic stress. Plant Sci 181:604–611

Couée I, Sulmon C, Gouesbet G, El Amrani A (2006) Involvement of soluble sugars in reactive oxygen species balance and responses to oxidative stress in plants. J Exp Bot 57:449–459

Couturier J, Jacquot J-P, Rouhier N (2009) Evolution and diversity of glutaredoxins in photosynthetic organisms. Cell Mol Life Sci 66:2539–2557

Culotta VC, Yang M, O'Halloran TV (2006) Activation of superoxide dismutases: putting the metal to the pedal. Biochim Biophys Acta 1763:747–758

Cuypers A, Smeets K, Ruytinx J, Opdenakker K, Keunen E, Remans T, Horemans N, Vanhoudt N, Van Sanden S, Van Belleghem F, Yvese G, Jana C, Jacoa V (2011) The cellular redox state as a modulator in cadmium and copper responses in *Arabidopsis thaliana* seedlings. J Plant Physiol 168:309–316

D'Autreaux B, Toledano MB (2007) ROS as signalling molecules: mechanisms that generate specificity in ROS homeostasis. Nat Rev Mol Cell Biol 8:813–824

Danh LT, Truong P, Mammucari R, Foster N (2011) Effect of calcium on growth performance and essential oil of vetiver grass (*Chrysopogon zizanioides*) grown on lead contaminated soils. Int J Phytoremediation 13:154–165

Debenest T, Silvestre J, Coste M, Pinelli E (2010) Effects of pesticides on freshwater diatoms. Rev Environ Contam Toxicol 203:87–103

Del Río LA (2011) Peroxisomes as a cellular source of reactive nitrogen species signal molecules. Arch Biochem Biophys 506:1–11

Deng X, Xia Y, Hu W, Zhang H, Shen Z (2010) Cadmium-induced oxidative damage and protective effects of N-acetyl-L-cysteine against cadmium toxicity in *Solanum nigrum* L. J Hazard Mater 180:722–729

Deponte M (2013) Glutathione catalysis and the reaction mechanisms of glutathione-dependent enzymes. Biochim Biophys Acta 1830:3217–3266

Dietz K-J (2003) Plant peroxiredoxins. Annu Rev Plant Biol 54:93–107

Dietz K-J (2011) Peroxiredoxins in plants and cyanobacteria. Antioxid Redox Signal 15:1129–1159

Djuika CF, Fiedler S, Schnölzer M, Sanchez C, Lanzer M, Deponte M (2013) *Plasmodium falciparum* antioxidant protein as a model enzyme for a special class of glutaredoxin/glutathione-dependent peroxiredoxins. Biochim Biophys Acta 1830:4073–4090

Dong C-J, Wang X-L, Shang Q-M (2011) Salicylic acid regulates sugar metabolism that confers tolerance to salinity stress in cucumber seedlings. Sci Hortic 129:629–636

Dubreuil-Maurizi C, Vitecek J, Marty L, Branciard L, Frettinger P, Wendehenne D, Meyer AJ, Mauch F, Poinssot B (2011) Glutathione deficiency of the arabidopsis mutant pad2-1 affects oxidative stress-related events, defense gene expression, and the hypersensitive response. Plant Physiol 157:2000–2012

Duquesnoy I, Champeau GM, Evray G, Ledoigt G, Piquet-Pissaloux A (2010) Enzymatic adaptations to arsenic-induced oxidative stress in *Zea mays* and genotoxic effect of arsenic in root tips of *Vicia faba* and *Zea mays*. C R Biol 333:814–824

Edreva A (2005) Generation and scavenging of reactive oxygen species in chloroplasts: a submolecular approach. Agric Ecosyst Environ 106:119–133

Ellis HR, Poole LB (1997) Roles for the two cysteine residues of AhpC in catalysis of peroxide reduction by alkyl hydroperoxide reductase from *Salmonella typhimurium*. Biochemistry 36:13349–13356

Evangelou MWH, Hockmann K, Pokharel R, Jakob A, Schulin R (2012) Accumulation of Sb, Pb, Cu, Zn and Cd by various plants species on two different relocated military shooting range soils. J Environ Manage 108:102–107

Fariduddin Q, Yusuf M, Hayat S, Ahmad A (2009) Effect of 28-homobrassinolide on antioxidant capacity and photosynthesis in *Brassica juncea* plants exposed to different levels of copper. Environ Exp Bot 66:418–424

Farmer EE, Mueller MJ (2013) ROS-Mediated lipid peroxidation and RES-activated signaling. Annu Rev Plant Biol 64:429–450

Foreman J, Demidchik V, Bothwell JHF, Mylona P, Miedema H, Torres MA, Linstead P, Costa S, Brownlee C, Jones JDG et al (2003) Reactive oxygen species produced by NADPH oxidase regulate plant cell growth. Nature 422:442–446

Foucault Y, Lévêque T, Xiong T, Schreck E, Austruy A, Shahid M, Dumat C (2013) Green manure plants for remediation of soils polluted by metals and metalloids: ecotoxicity and human bioavailability assessment. Chemosphere. doi:10.1016/j.chemosphere.2013.07.040

Foyer CH, Noctor G (2003) Redox sensing and signalling associated with reactive oxygen in chloroplasts, peroxisomes and mitochondria. Physiol Plant 119:355–364

Foyer CH, Noctor G (2005) Redox homeostasis and antioxidant signaling: a metabolic interface between stress perception and physiological responses. Plant Cell 17:1866–1875

Foyer CH, Noctor G (2009) Redox regulation in photosynthetic organisms: signaling, acclimation, and practical implications. Antioxid Redox Signal 11:861–905

Foyer CH, Noctor G (2012) Managing the cellular redox hub in photosynthetic organisms. Plant Cell Environ 35:199–201

Freeman JL, Quinn CF, Marcus MA, Fakra S, Pilon-Smits EAH (2006) Selenium-tolerant diamondback moth disarms hyperaccumulator plant defense. Curr Biol 16:2181–2192

Gadjev I, Stone JM, Gechev TS (2008) Programmed cell death in plants: new insights into redox regulation and the role of hydrogen peroxide. Int Rev Cell Mol Biol 270:87–144

Gao S, Ou-yang C, Tang L, Zhu J-q, Xu Y, S-hua W, Chen F (2010) Growth and antioxidant responses in *Jatropha curcas* seedling exposed to mercury toxicity. J Hazard Mater 182:591–597

Garcia JS, Gratão PL, Azevedo RA, Arruda MAZ (2006) Metal contamination effects on sunflower (*Helianthus annuus* L. growth and protein expression in leaves during development. J Agric Food Chem 54:8623–8630

Gastaldo J, Viau M, Bouchot M, Joubert A, Charvet A-M, Foray N (2008) Induction and repair rate of DNA damage: a unified model for describing effects of external and internal irradiation and contamination with heavy metals. J Theor Biol 251:68–81

Gémes K, Poór P, Horváth E, Kolbert Z, Szopkó D, Szepesi Á, Tari I (2011) Cross-talk between salicylic acid and NaCl-generated reactive oxygen species and nitric oxide in tomato during acclimation to high salinity. Physiol Plant 142:179–192

Giaccio L, Cicchella D, De Vivo B, Lombardi G, De Rosa M (2012) Does heavy metals pollution affects semen quality in men? A case of study in the metropolitan area of Naples (Italy). J Geochem Explor 112:218–225

Gichner T, Patková Z, Száková J, Demnerová K (2006) Toxicity and DNA damage in tobacco and potato plants growing on soil polluted with heavy metals. Ecotoxicol Environ Saf 65:420–426

Gill SS, Tuteja N (2010) Reactive oxygen species and antioxidant machinery in abiotic stress tolerance in crop plants. Plant Physiol Biochem 48:909–930

Ginn BR, Szymanowski JS, Fein JB (2008) Metal and proton binding onto the roots of *Fescue rubra*. Chem Geol 253:130–135

Gisbert C, Ros R, De Haro A, Walker DJ, Pilar Bernal M, Serrano R, Navarro-Aviñó J (2003) A plant genetically modified that accumulates Pb is especially promising for phytoremediation. Biochem Biophys Res Commun 303:440–445

Gonnelli C, Galardi F, Gabbrielli R (2001) Nickel and copper tolerance and toxicity in three Tuscan populations of *Silene paradoxa*. Physiol Plant 113:507–514

Gopal R, Rizvi AH (2008) Excess lead alters growth, metabolism and translocation of certain nutrients in radish. Chemosphere 70:1539–1544

Groppa MD, Rosales EP, Iannone MF, Benavides MP (2008) Nitric oxide, polyamines and Cd-induced phytotoxicity in wheat roots. Phytochemistry 69:2609–2615

Grotz N, Guerinot ML (2006) Molecular aspects of Cu, Fe and Zn homeostasis in plants. Biochim Biophys Acta 1763:595–608

Grover P, Rekhadevi PV, Danadevi K, Vuyyuri SB, Mahboob M, Rahman MF (2010) Genotoxicity evaluation in workers occupationally exposed to lead. Int J Hyg Environ Health 213:99–106

Guan Z, Chai T, Zhang Y, Xu J, Wei W (2009) Enhancement of Cd tolerance in transgenic tobacco plants overexpressing a Cd-induced catalase cDNA. Chemosphere 76:623–630

Guan-fu F (2011) Changes of oxidative stress and soluble sugar in anthers involve in rice pollen abortion under drought stress. Agric Sci China 10:1016–1025

Guo TR, Zhang GP, Zhang YH (2007) Physiological changes in barley plants under combined toxicity of aluminum, copper and cadmium. Colloids Surf B: Biointerfaces 57:182–188

Gupta M, Sarin NB (2009) Heavy metal induced DNA changes in aquatic macrophytes: random amplified polymorphic DNA analysis and identification of sequence characterized amplified region marker. J Environ Sci (China) 21:686–690

Gupta AK, Sinha S (2009) Antioxidant response in sesame plants grown on industrially contaminated soil: effect on oil yield and tolerance to lipid peroxidation. Bioresour Technol 100:179–185

Gupta DK, Nicoloso FT, Schetinger MRC, Rossato LV, Pereira LB, Castro GY, Srivastava S, Tripathi RD (2009) Antioxidant defense mechanism in hydroponically grown *Zea mays* seedlings under moderate lead stress. J Hazard Mater 172:479–484

Gupta DK, Huang HG, Yang XE, Razafindrabe BHN, Inouhe M (2010) The detoxification of lead in *Sedum alfredii* H. is not related to phytochelatins but the glutathione. J Hazard Mater 177:437–444

Hajeb P, Jinap S, Ismail A, Mahyudin NA (2011) Mercury pollution in Malaysia. Rev Environ Contam Toxicol 220:45–66

Hall JL, Williams LE (2003) Transition metal transporters in plants. J Exp Bot 54:2601–2613

Halliwell B, Gutteridge JMC (1999) Free radicals in biology and medicine, 3rd edn. Clarendon, Oxford

Hanikenne M, Krämer U, Demoulin V, Baurain D (2005) A Comparative Inventory of Metal Transporters in the green alga *Chlamydomonas reinhardtii* and the red alga *Cyanidioschizon merolae*. Plant Physiol 137:428–446

Hao F, Wang X, Chen J (2006) Involvement of plasma-membrane NADPH oxidase in nickel-induced oxidative stress in roots of wheat seedlings. Plant Sci 170:151–158

Harris GK, Shi X (2003) Signaling by carcinogenic metals and metal-induced reactive oxygen species. Mutat Res 533:183–200

Hassan Z, Aarts MGM (2011) Opportunities and feasibilities for biotechnological improvement of Zn, Cd or Ni tolerance and accumulation in plants. Environ Exp Bot 72:53–63

He H-Y, He L-F, Gu M-H, Li X-F (2012) Nitric oxide improves aluminum tolerance by regulating hormonal equilibrium in the root apices of rye and wheat. Plant Sci 183:123–130

He J, Qin J, Long L, Ma Y, Li H, Li K, Jiang X, Liu T, Polle A, Liang Z et al (2011) Net cadmium flux and accumulation reveal tissue-specific oxidative stress and detoxification in Populus × canescens. Physiol Plant 143:50–63

Hermes-Lima M (2005) Oxygen in biology and biochemistry: role of free radicals. In: Storey KB (ed) Functional metabolism: regulation and adaptation. Wiley, Hoboken, pp 319–368

Hirano T, Tamae K (2010) Heavy metal-induced oxidative DNA damage in earthworms: a review. Appl Environ Soil Sci 2010: Article ID 726946, 7 p

Hirata A, Corcoran GB, Hirata F (2010) Carcinogenic heavy metals replace Ca^{2+} for DNA binding and annealing activities of mono-ubiquitinated annexin A1 homodimer. Toxicol Appl Pharmacol 248:45–51

Hirata A, Corcoran GB, Hirata F (2011) Carcinogenic heavy metals, As^{3+} and Cr^{6+}, increase affinity of nuclear mono-ubiquitinated annexin A1 for DNA containing 8-oxo-guanosine, and promote translesion DNA synthesis. Toxicol Appl Pharmacol 252:159–164

Hu L, McBride MB, Cheng H, Wu J, Shi J, Xu J, Wu L (2011) Root-induced changes to cadmium speciation in the rhizosphere of two rice (*Oryza sativa* L.) genotypes. Environ Res 111:356–361

Hu R, Sun K, Su X, Pan Y-X, Zhang Y-F, Wang X-P (2012) Physiological responses and tolerance mechanisms to Pb in two xerophils: *Salsola passerina* Bunge and *Chenopodium album* L. J Hazard Mater 205–206:131–138

Huang G-Y, Wang Y-S (2010) Physiological and biochemical responses in the leaves of two mangrove plant seedlings (*Kandelia candel* and *Bruguiera gymnorrhiza*) exposed to multiple heavy metals. J Hazard Mater 182:848–854

Huang H, Li T, Tian S, Gupta DK, Zhang X, Yang X-E (2008) Role of EDTA in alleviating lead toxicity in accumulator species of *Sedum alfredii* H. Bioresour Technol 99:6088–6096

Hummel M, Rahmani F, Smeekens S, Hanson J (2009) Sucrose-mediated translational control. Ann Bot 104:1–7

Hunt PR, Olejnik N, Robert RS (2012) Toxicity ranking of heavy metals with screening method using adult *Caenorhabditis elegans* and propidium iodide replicates toxicity ranking in rat. Food Chem Toxicol 50:3280–90. doi:10.1016/j.fct.2012.06.051

Israr M, Jewell A, Kumar D, Sahi SV (2011) Interactive effects of lead, copper, nickel and zinc on growth, metal uptake and antioxidative metabolism of *Sesbania drummondii*. J Hazard Mater 186:1520–1526

Islam E, Liu D, Li T, Yang X, Jin X, Mahmood Q, Tian S, Li J (2008) Effect of Pb toxicity on leaf growth, physiology and ultrastructure in the two ecotypes of *Elsholtzia argyi*. J Hazard Mater 154:914–926

Ito H, Iwabuchi M, Ogawa K (2003) The sugar-metabolic enzymes aldolase and triose-phosphate isomerase are targets of glutathionylation in *Arabidopsis thaliana*: detection using biotinylated glutathione. Plant Cell Physiol 44:655–660

Jasinski M, Sudre D, Schansker G, Schellenberg M, Constant S, Martinoia E, Bovet L (2008) AtOSA1, a member of the Abc1-like family, as a new factor in cadmium and oxidative stress response. Plant Physiol 147:719–731

Jaspers P, Kangasjärvi J (2010) Reactive oxygen species in abiotic stress signaling. Physiol Plant 138:405–413

Jia Y, Ju X, Liao S, Song Z, Li Z (2011) Phytochelatin synthesis in response to elevated CO_2 under cadmium stress in *Lolium perenne* L. J Plant Physiol 168:1723–1728

Jiang W, Liu D (2010) Pb-induced cellular defense system in the root meristematic cells of *Allium sativum* L. BMC Plant Biol 10:40

Johnson FM (1998) The genetic effects of environmental lead. Mutat Res 410:123–140

Jonak C, Nakagami H, Hirt H (2004) Heavy metal stress. Activation of distinct mitogen-activated protein kinase pathways by copper and cadmium. Plant Physiol 136:3276–3283

Jones GC, Corin KC, van Hille RP, Harrison STL (2011) The generation of toxic reactive oxygen species (ROS) from mechanically activated sulphide concentrates and its effect on thermophilic bioleaching. Miner Eng 24:1198–1208

Juan K, Hong-mei W, Chang-hai J, Hai-yan X (2010) Changes of reactive oxygen species and related enzymes in mitochondria respiratory metabolism during the ripening of peach fruit. Agric Sci China 9:138–146

Kafel A, Nadgórska-Socha A, Gospodarek J, Babczyńska A, Skowronek M, Kandziora M, Rozpedek K (2010) The effects of *Aphis fabae* infestation on the antioxidant response and heavy metal content in field grown *Philadelphus coronarius* plants. Sci Total Environ 408:1111–1119

Kanwar MK, Bhardwaj R, Arora P, Chowdhary SP, Sharma P, Kumar S (2012) Plant steroid hormones produced under Ni stress are involved in the regulation of metal uptake and oxidative stress in *Brassica juncea* L. Chemosphere 86:41–49

Karuppanapandian T, Moon J, Kim C, Manoharan K, Kim W (2011a) Reactive oxygen species in plants: their generation, signal transduction, and scavenging mechanisms. Aust J Crop Sci 5:709–725

Karuppanapandian T, Wang HW, Prabakaran N, Jeyalakshmi K, Kwon M, Manoharan K, Kim W (2011b) 2,4-dichlorophenoxyacetic acid-induced leaf senescence in mung bean (*Vigna radiata* L. Wilczek) and senescence inhibition by co-treatment with silver nanoparticles. Plant Physiol Biochem 49:168–177

Kehrer JP (2000) The Haber–Weiss reaction and mechanisms of toxicity. Toxicology 149:43–50

Kerchev PI, Pellny TK, Vivancos PD, Kiddle G, Hedden P, Driscoll S, Vanacker H, Verrier P, Hancock RD, Foyer CH (2011) The transcription factor ABI4 is required for the ascorbic acid–dependent regulation of growth and regulation of jasmonate-dependent defense signaling pathways in arabidopsis. Plant Cell 23:3319–3334

Kerin EJ, Lin HK (2010) Fugitive dust and human exposure to heavy metals around the red dog mine. Rev Environ Contam Toxicol 206:49–63

Khatun S, Ali MB, Hahn E-J, Paek K-Y (2008) Copper toxicity in *Withania somnifera*: growth and antioxidant enzymes responses of in vitro grown plants. Environ Exp Bot 64:279–285

Kim Y-H, Lee H-S, Kwak S-S (2010) Differential responses of sweet potato peroxidases to heavy metals. Chemosphere 81:79–85

Klatte M, Schuler M, Wirtz M, Fink-Straube C, Hell R, Bauer P (2009) The analysis of *Arabidopsis nicotianamine* synthase mutants reveals functions for nicotianamine in seed iron loading and iron deficiency responses. Plant Physiol 150:257–271

Kopittke PM, Asher CJ, Blamey FPC, Menzies NW (2007) Toxic effects of Pb^{2+} on the growth and mineral nutrition of signal grass (*Brachiaria decumbens*) and Rhodes grass (*Chloris gayana*). Plant Soil 300:127–136

Kopyra M, Gwóźdź EA (2003) Nitric oxide stimulates seed germination and counteracts the inhibitory effect of heavy metals and salinity on root growth of *Lupinus luteus*. Plant Physiol Biochem 41:1011–1017

Körpe DA, Aras S (2011) Evaluation of copper-induced stress on eggplant (*Solanum melongena* L.) seedlings at the molecular and population levels by use of various biomarkers. Mutat Res 719:29–34

Koutsogiannaki S, Evangelinos N, Koliakos G, Kaloyianni M (2006) Cytotoxic mechanisms of Zn^{2+} and Cd^{2+} involve Na^+/H^+ exchanger (NHE) activation by ROS. Aquat Toxicol 78:315–324

Kovac J, Klejdus B, Kadukova J, Backor M (2009) Physiology of *Matricaria chamomilla* exposed to nickel excess. Ecotoxicol Environ Saf 72:603–609

Kovacic P, Somanathan R (2010) Dermal toxicity and environmental contamination: electron transfer, reactive oxygen species, oxidative stress, cell signaling, and protection by antioxidants. Rev Environ Contam Toxicol 203:119–138

Kováčik J, Klejdus B, Hedbavny J, Bačkor M (2010) Effect of copper and salicylic acid on phenolic metabolites and free amino acids in *Scenedesmus quadricauda* (Chlorophyceae). Plant Sci 178:307–311

Kovalchuk I, Titov V, Hohn B, Kovalchuk O (2005) Transcriptome profiling reveals similarities and differences in plant responses to cadmium and lead. Mutat Res 570:149–161

Kranner I, Roach T, Beckett RP, Whitaker C, Minibayeva FV (2010) Extracellular production of reactive oxygen species during seed germination and early seedling growth in *Pisum sativum*. J Plant Physiol 167:805–811

Krzesłowska M, Lenartowska M, Mellerowicz EJ, Samardakiewicz S, Woźny A (2009) Pectinous cell wall thickenings formation—A response of moss protonemata cells to lead. Environ Exp Bot 65:119–131

Krzesłowska M, Lenartowska M, Samardakiewicz S, Bilski H, Woźny A (2010) Lead deposited in the cell wall of Funaria hygrometrica protonemata is not stable–a remobilization can occur. Environ Pollut 158:325–338

Kumar M, Bijo AJ, Baghel RS, Reddy CRK, Jha B (2012) Selenium and spermine alleviate cadmium induced toxicity in the red seaweed *Gracilaria dura* by regulating antioxidants and DNA methylation. Plant Physiol Biochem 51:129–138

Küpper H, Lombi E, Zhao FJ, McGrath SP (2000) Cellular compartmentation of cadmium and zinc in relation to other elements in the hyperaccumulator *Arabidopsis halleri*. Planta 212:75–84

Labra M, Gianazza E, Waitt R, Eberini I, Sozzi A, Regondi S, Grassi F, Agradi E (2006) *Zea mays* L. protein changes in response to potassium dichromate treatments. Chemosphere 62:1234–1244

Lea PJ, Azevedo RA (2007) Nitrogen use efficiency. 2. Amino acid metabolism. Ann Appl Biol 151:269–275

Lee S, Moon JS, Ko T-S, Petros D, Goldsbrough PB, Korban SS (2003) Overexpression of arabidopsis phytochelatin synthase paradoxically leads to hypersensitivity to cadmium stress. Plant Physiol 131:656–663

Lee J-C, Son Y-O, Pratheeshkumar P, Shi X (2012) Oxidative stress and metal carcinogenesis. Free Radic Biol Med 53:742–57. doi:10.1016/j.freeradbiomed.2012.06.002

Lehner A, Mamadou N, Poels P, Côme D, Bailly C, Corbineau F (2008) Changes in soluble carbohydrates, lipid peroxidation and antioxidant enzyme activities in the embryo during ageing in wheat grains. J Cereal Sci 47:555–565

Lemaire SD, Collin V, Keryer E, Issakidis-Bourguet E, Lavergne D, Miginiac-Maslow M (2003) *Chlamydomonas reinhardtii*: a model organism for the study of the thioredoxin family. Plant Physiol Biochem 41:513–521

Li S, Zachgo S (2009) Glutaredoxins in development and stress responses of plants. In: Jacquot J-P (ed) Advances in botanical research. Academic, Boston, pp 333–361, Chapter 11

Lin A-J, Zhang X-H, Chen M-M, Cao Q (2007) Oxidative stress and DNA damages induced by cadmium accumulation. J Environ Sci (China) 19:596–602

Liu F, Pang SJ (2010) Stress tolerance and antioxidant enzymatic activities in the metabolisms of the reactive oxygen species in two intertidal red algae *Grateloupia turuturu* and *Palmaria palmata*. J Exp Mar Biol Ecol 382:82–87

Liu H, Liao B, Lu S (2004) Toxicity of surfactant, acid rain and Cd^{2+} combined pollution to the nucleus of *Vicia faba* root tip cells. Chin J Appl Ecol 15:493–496

Liu D, Li T-Q, Jin X-F, Yang X-E, Islam E, Mahmood Q (2008) Lead induced changes in the growth and antioxidant metabolism of the lead accumulating and non-accumulating ecotypes of *Sedum alfredii*. J Integr Plant Biol 50:129–140

Liu T, Liu S, Guan H, Ma L, Chen Z, Gu H, Qu L-J (2009) Transcriptional profiling of Arabidopsis seedlings in response to heavy metal lead (Pb). Environ Exp Bot 67:377–386

Liu N, Lin Z-F, Lin G-Z, Song L-Y, Chen S-W, Mo H, Peng C-L (2010) Lead and cadmium induced alterations of cellular functions in leaves of *Alocasia macrorrhiza* L. Schott. Ecotoxicol Environ Saf 73:1238–1245

Lomonte C, Sgherri C, Baker AJM, Kolev SD, Navari-Izzo F (2010) Antioxidative response of Atriplex codonocarpa to mercury. Environ Exp Bot 69:9–16

Louriño-Cabana B, Lesven L, Charriau A, Billon G, Ouddane B, Boughriet A (2011) Potential risks of metal toxicity in contaminated sediments of Deûle river in northern France. J Hazard Mater 186:2129–2137

Luo X-S, Ding J, Xu B, Wang Y-J, Li H-B, Yu S (2012) Incorporating bioaccessibility into human health risk assessments of heavy metals in urban park soils. Sci Total Environ 424:88–96

Luque-Garcia JL, Cabezas-Sanchez P, Camara C (2011) Proteomics as a tool for examining the toxicity of heavy metals. Trends Anal Chem 30:703–716

Lushchak VI (2011) Environmentally induced oxidative stress in aquatic animals. Aquat Toxicol 101:13–30

Lyubenova L, Schröder P (2011) Plants for waste water treatment–effects of heavy metals on the detoxification system of Typha latifolia. Bioresour Technol 102:996–1004

Maestri E, Marmiroli M, Visioli G, Marmiroli N (2010) Metal tolerance and hyperaccumulation: costs and trade-offs between traits and environment. Environ Exp Bot 68:1–13

Mallick S, Sinam G, Kumar Mishra R, Sinha S (2010) Interactive effects of Cr and Fe treatments on plants growth, nutrition and oxidative status in *Zea mays* L. Ecotoxicol Environ Saf 73:987–995

Mano J (2012) Reactive carbonyl species: their production from lipid peroxides, action in environmental stress, and the detoxification mechanism. Plant Physiol Biochem 59:90–97

Marcato-Romain C-E, Guiresse M, Cecchi M, Cotelle S, Pinelli E (2009a) New direct contact approach to evaluate soil genotoxicity using the *Vicia faba* micronucleus test. Chemosphere 77:345–350

Marcato-Romain C-E, Pinelli E, Pourrut B, Silvestre J, Guiresse M (2009b) Assessment of the genotoxicity of Cu and Zn in raw and anaerobically digested slurry with the *Vicia faba* micronucleus test. Mutat Res 672:113–118

Marnett LJ (1987) Peroxyl free radicals: potential mediators of tumor initiation and promotion. Carcinogenesis 8:1365–1373

Márquez-García B, Pérez-López R, Ruíz-Chancho MJ, López-Sánchez JF, Rubio R, Abreu MM, Nieto JM, Córdoba F (2012) Arsenic speciation in soils and *Erica andevalensis* Cabezudo & Rivera and *Erica australis* L. from São Domingos Mine area, Portugal. J Geochem Explor 119–120:51–59.

Márquez-García B, Horemans N, Cuypers A, Guisez Y, Córdoba F (2011) Antioxidants in *Erica andevalensis*: a comparative study between wild plants and cadmium-exposed plants under controlled conditions. Plant Physiol Biochem 49:110–115

Martínez Domínguez D, Torronteras Santiago R, Córdoba García F (2009) Modulation of the antioxidative response of *Spartina densiflora* against iron exposure. Physiol Plant 136:169–179

Martínez Domínguez D, Córdoba García F, Canalejo Raya A, Torronteras Santiago R (2010) Cadmium-induced oxidative stress and the response of the antioxidative defense system in *Spartina densiflora*. Physiol Plant 139:289–302

Martínez-Fernández D, Walker DJ, Romero-Espinar P, Flores P, del Río JA (2011) Physiological responses of *Bituminaria bituminosa* to heavy metals. J Plant Physiol 168:2206–2211

Martínez-Peñalver A, Graña E, Reigosa MJ, Sánchez-Moreiras AM (2012) The early response of *Arabidopsis thaliana* to cadmium- and copper-induced stress. Environ Exp Bot 78:1–9

Martinoia E, Maeshima M, Neuhaus HE (2007) Vacuolar transporters and their essential role in plant metabolism. J Exp Bot 58:83–102

Matamoros MA, Loscos J, Dietz K-J, Aparicio-Tejo PM, Becana M (2010) Function of antioxidant enzymes and metabolites during maturation of pea fruits. J Exp Bot 61:87–97

McCord JM, Fridovich I (1969) Superoxide dismutase: an enzymic function for erythrocuprein (hemocuprein). J Biol Chem 244:6049–6055

Meyer AJ (2008) The integration of glutathione homeostasis and redox signaling. J Plant Physiol 165:1390–1403

Meyers DER, Auchterlonie GJ, Webb RI, Wood B (2008) Uptake and localisation of lead in the root system of *Brassica juncea*. Environ Pollut 153:323–332

Michelet L, Zaffagnini M, Massot V, Keryer E, Vanacker H, Miginiac-Maslow M, Issakidis-Bourguet E, Lemaire SD (2006) Thioredoxins, glutaredoxins, and glutathionylation: new crosstalks to explore. Photosynth Res 89:225–245

Mingorance MD, Leidi EO, Valdés V, Oliv SR (2012) Evaluation of lead toxicity in *Erica andevalensis* as an alternative species for revegetation of contaminated soils. Int J Phytoremediation 14:174–185

Minibayeva F, Dmitrieva S, Ponomareva A, Ryabovol V (2012) Oxidative stress-induced autophagy in plants: the role of mitochondria. Plant Physiol Biochem. doi:10.1016/j.plaphy.2012.02.013

Mishra S, Srivastava S, Tripathi RD, Kumar R, Seth CS, Gupta DK (2006) Lead detoxification by coontail (*Ceratophyllum demersum* L.) involves induction of phytochelatins and antioxidant system in response to its accumulation. Chemosphere 65:1027–1039

Mittler R (2002) Oxidative stress, antioxidants and stress tolerance. Trends Plant Sci 7:405–410

Mittler R, Vanderauwera S, Gollery M, Van Breusegem F (2004) Reactive oxygen gene network of plants. Trends Plant Sci 9:490–498

Møller IM, Jensen PE, Hansson A (2007) Oxidative modifications to cellular components in plants. Annu Rev Plant Biol 58:459–481

Morina F, Jovanovic L, Mojovic M, Vidovic M, Pankovic D, Veljovic Jovanovic S (2010) Zinc-induced oxidative stress in *Verbascum thapsus* is caused by an accumulation of reactive oxygen species and quinhydrone in the cell wall. Physiol Plant 140:209–224

Mou D, Yao Y, Yang Y, Zhang Y, Tian C, Achal V (2011) Plant high tolerance to excess manganese related with root growth, manganese distribution and antioxidative enzyme activity in three grape cultivars. Ecotoxicol Environ Saf 74:776–786

Mukherjee A, Das D, Kumar Mondal S, Biswas R, Kumar Das T, Boujedaini N, Khuda-Bukhsh AR (2010) Tolerance of arsenate-induced stress in *Aspergillus niger*, a possible candidate for bioremediation. Ecotoxicol Environ Saf 73:172–182

Na G, Salt DE (2010) The role of sulfur assimilation and sulfur-containing compounds in trace element homeostasis in plants. Environ Exp Bot 72:18–25

Nanthi S, Bolan GC (2012) Microbial transformation of trace elements in soils in relation to bioavailability and remediation. Rev Environ Contam Toxicol 225:1–56

Nasim SA, Dhir B (2010) Heavy metals alter the potency of medicinal plants. Rev Environ Contam Toxicol 203:139–149

Nayek S, Gupta S, Saha RN (2010) Metal accumulation and its effects in relation to biochemical response of vegetables irrigated with metal contaminated water and wastewater. J Hazard Mater 178:588–595

Nehnevajova E, Lyubenova L, Herzig R, Schröder P, Schwitzguébel JP, Schmülling T (2012) Metal accumulation and response of antioxidant enzymes in seedlings and adult sunflower mutants with improved metal removal traits on a metal-contaminated soil. Environ Exp Bot 76:39–48

Nguyen GN, Hailstones DL, Wilkes M, Sutton BG (2010) Drought stress: role of carbohydrate metabolism in drought-induced male sterility in rice anthers. J Agron Crop Sci 196: 346–357

Nishiyama Y, Allakhverdiev SI, Murata N (2011) Protein synthesis is the primary target of reactive oxygen species in the photoinhibition of photosystem II. Physiol Plant 142:35–46

Noctor G, Arisi AC, Jouanin L, Kunert KJ, Rennenberg H, Foyer CH (1998) Glutathione: biosynthesis, metabolism and relationship to stress tolerance in transformed plants. J Exp Bot 49:623–647

Noctor G, Mhamdi A, Chaouch S, Han Y, Neukermans J, Marquez-Garcia B, Queval G, Foyer CH (2012) Glutathione in plants: an integrated overview. Plant Cell Environ 35:454–484

Oda K, Otani M, Uraguchi S, Akihiro T, Fujiwara T (2011) Rice ABCG43 is Cd inducible and confers Cd tolerance on yeast. Biosci Biotechnol Biochem 75:1211–1213

Ogawa S, Yoshidomi T, Yoshimura E (2011) Cadmium(II)-stimulated enzyme activation of *Arabidopsis thaliana* phytochelatin synthase I. J Inorg Biochem 105:111–117

Oliveira SCB, Corduneanu O, Oliveira-Brett AM (2008) In situ evaluation of heavy metal-DNA interactions using an electrochemical DNA biosensor. Bioelectrochemistry 72:53–58

Olmos E, Martínez-Solano JR, Piqueras A, Hellín E (2003) Early steps in the oxidative burst induced by cadmium in cultured tobacco cells (BY-2 line). J Exp Bot 54:291–301

Opdenakker K, Remans T, Keunen E, Vangronsveld J, Cuypers A (2012) Exposure of *Arabidopsis thaliana* to Cd or Cu excess leads to oxidative stress mediated alterations in MAPKinase transcript levels. Environ Exp Bot 83:53–61

Oracz K, El-Maarouf-Bouteau H, Kranner I, Bogatek R, Corbineau F, Bailly C (2009) The mechanisms involved in seed dormancy alleviation by hydrogen cyanide unravel the role of reactive oxygen species as key factors of cellular signaling during germination. Plant Physiol 150:494–505

Overmyer K, Brosché M, Kangasjärvi J (2003) Reactive oxygen species and hormonal control of cell death. Trends Plant Sci 8:335–342

Pál M, Horváth E, Janda T, Páldi E, Szalai G (2005) Cadmium stimulates the accumulation of salicylic acid and its putative precursors in maize (*Zea mays*) plants. Physiol Plant 125:356–364

Pandey SP, Somssich IE (2009) The role of WRKY transcription factors in plant immunity. Plant Physiol 150:1648–1655

Park Y, Moon Y, Ryoo J, Kim N, Cho H, Ahn JH (2012) Identification of the minimal region in lipase ABC transporter recognition domain of *Pseudomonas fluorescens* for secretion and fluorescence of green fluorescent protein. Microb Cell Fact 11:60

Pena LB, Pasquini LA, Tomaro ML, Gallego SM (2006) Proteolytic system in sunflower (*Helianthus annuus* L.) leaves under cadmium stress. Plant Sci 171:531–537

Pena LB, Pasquini LA, Tomaro ML, Gallego SM (2007) 20S proteasome and accumulation of oxidized and ubiquitinated proteins in maize leaves subjected to cadmium stress. Phytochemistry 68:1139–1146

Pena LB, Zawoznik MS, Tomaro ML, Gallego SM (2008) Heavy metals effects on proteolytic system in sunflower leaves. Chemosphere 72:741–746

Piotrowska A, Bajguz A, Godlewska-Zylkiewicz B, Czerpak R, Kaminska M (2009) Jasmonic acid as modulator of lead toxicity in aquatic plant *Wolffia arrhiza* (Lemnaceae). Environ Exp Bot 66:507–513

Piotrowska-Niczyporuk A, Bajguz A, Zambrzycka E, Godlewska-Żyłkiewicz B (2012) Phytohormones as regulators of heavy metal biosorption and toxicity in green alga *Chlorella vulgaris* (Chlorophyceae). Plant Physiol Biochem 52:52–65

Pitzschke A, Hirt H (2006) Mitogen-activated protein kinases and reactive oxygen species signaling in plants. Plant Physiol 141:351–356

Pócsi I, Prade RA, Penninckx MJ (2004) Glutathione, altruistic metabolite in fungi. Adv Microb Physiol 49:1–76

Poole LB, Karplus PA, Claiborne A (2004) Protein sulfenic acids in redox signaling. Annu Rev Pharmacol Toxicol 44:325–347

Porta H, Rocha-Sosa M (2002) Plant Lipoxygenases. Physiological and Molecular Features. Plant Physiol 130:15–21

Potocký M, Pejchar P, Gutkowska M, Jiménez-Quesada MJ, Potocká A, Alché Jde D, Kost B, Zárský V (2012) NADPH oxidase activity in pollen tubes is affected by calcium ions, signaling phospholipids and Rac/Rop GTPases. J Plant Physiol 169(16):1654–63. doi:10.1016/j.jplph.2012.05.014

Pourrut B, Perchet G, Silvestre J, Cecchi M, Guiresse M, Pinelli E (2008) Potential role of NADPH-oxidase in early steps of lead-induced oxidative burst in *Vicia faba* roots. J Plant Physiol 165:571–579

Pourrut B, Jean S, Silvestre J, Pinelli E (2011a) Lead-induced DNA damage in *Vicia faba* root cells: potential involvement of oxidative stress. Mutat Res 726:123–128

Pourrut B, Pohu AL, Pruvot C, Garçon G, Verdin A, Waterlot C, Bidar G, Shirali P, Douay F (2011b) Assessment of fly ash-aided phytostabilisation of highly contaminated soils after an 8-year field trial Part 2. Influence on plants. Sci Total Environ 409:4504–4510

Pourrut B, Shahid M, Dumat C, Winterton P, Pinelli E (2011c) Lead uptake, toxicity and detoxification in plants. Rev Environ Contam Toxicol 213:113–136

Pourrut B, Shahid M, Douay F, Dumat C, Pinelli E (2013) Molecular mechanisms involved in lead uptake, toxicity and detoxification in higher plants. In: Corpas FJ, Palma JM, Gupta DK (eds) Heavy metal stress in plants. Springer, Berlin, pp 121–147

Poynton RA, Hampton MB (2013) Peroxiredoxins as biomarkers of oxidative stress. Biochim Biophys Acta 1840:906–12. doi:10.1016/j.bbagen.2013.08.001

Prasad TK (1996) Mechanisms of chilling-induced oxidative stress injury and tolerance in developing maize seedlings: changes in antioxidant system, oxidation of proteins and lipids, and protease activities. Plant J 10:1017–1026

Prévéral S, Gayet L, Moldes C, Hoffmann J, Mounicou S, Gruet A, Reynaud F, Lobinski R, Verbavatz J-M, Vavasseur A et al (2009) A common highly conserved cadmium detoxification mechanism from bacteria to humans: heavy metal tolerance conferred by the ATP-binding cassette (ABC) transporter SpHMT1 requires glutathione but not metal-chelating phytochelatin peptides. J Biol Chem 284:4936–4943

Price J, Laxmi A, St Martin SK, Jang J-C (2004) Global transcription profiling reveals multiple sugar signal transduction mechanisms in Arabidopsis. Plant Cell 16:2128–2150

Probst A, Liu H, Fanjul M, Liao B, Hollande E (2009) Response of *Vicia faba* L. to metal toxicity on mine tailing substrate: geochemical and morphological changes in leaf and root. Environ Exp Bot 66:297–308

Pucciariello C, Banti V, Perata P (2012) ROS signaling as common element in low oxygen and heat stresses. Plant Physiol Biochem 59:3–10. doi:10.1016/j.plaphy.2012.02.016

Qiu R-L, Zhao X, Tang Y-T, Yu F-M, Hu P-J (2008) Antioxidative response to Cd in a newly discovered cadmium hyperaccumulator, *Arabis paniculata* F. Chemosphere 74:6–12

Radić S, Babić M, Skobić D, Roje V, Pevalek-Kozlina B (2010) Ecotoxicological effects of aluminum and zinc on growth and antioxidants in *Lemna minor* L. Ecotoxicol Environ Saf 73:336–342

Radić S, Stipaničev D, Cvjetko P, Marijanović Rajčić M, Sirac S, Pevalek-Kozlina B, Pavlica M (2011) Duckweed Lemna minor as a tool for testing toxicity and genotoxicity of surface waters. Ecotoxicol Environ Saf 74:182–187

Radwan MA, El-Gendy KS, Gad AF (2010) Biomarkers of oxidative stress in the land snail, *Theba pisana* for assessing ecotoxicological effects of urban metal pollution. Chemosphere 79:40–46

Rai MK, Kalia RK, Singh R, Gangola MP, Dhawan AK (2011) Developing stress tolerant plants through in vitro selection—an overview of the recent progress. Environ Exp Bot 71:89–98

Randall LM, Ferrer-Sueta G, Denicola A (2013) Peroxiredoxins as preferential targets in H_2O_2-induced signaling. Methods Enzymol 527:41–63

Rascio N, Navari-Izzo F (2011) Heavy metal hyperaccumulating plants: how and why do they do it? And what makes them so interesting? Plant Sci 180:169–181

Rawlings ND (2004) MEROPS: the peptidase database. Nucleic Acids Res 32:160–164

Rea PA (2012) Phytochelatin synthase: of a protease a peptide polymerase made. Physiol Plant 145:154–164

Requejo R, Tena M (2005) Proteome analysis of maize roots reveals that oxidative stress is a main contributing factor to plant arsenic toxicity. Phytochemistry 66:1519–1528

Roach T, Beckett RP, Minibayeva FV, Colville L, Whitaker C, Chen H, Bailly C, Kranner I (2010) Extracellular superoxide production, viability and redox poise in response to desiccation in recalcitrant *Castanea sativa* seeds. Plant Cell Environ 33:59–75

Robinson BH, Lombi E, Zhao FJ, McGrath SP (2003) Uptake and distribution of nickel and other metals in the hyperaccumulator *Berkheya coddii*. New Phytol 158:279–285

Rodríguez-Serrano M, Romero-Puertas MC, Zabalza A, Corpas FJ, Gómez M, Del Río LA, Sandalio LM (2006) Cadmium effect on oxidative metabolism of pea (*Pisum sativum* L.) roots. Imaging of reactive oxygen species and nitric oxide accumulation in vivo. Plant Cell Environ 29:1532–1544

Rodríguez-Serrano M, Romero-Puertas MC, Sparkes I, Hawes C, del Río LA, Sandalio LM (2009) Peroxisome dynamics in Arabidopsis plants under oxidative stress induced by cadmium. Free Radic Biol Med 47:1632–1639

Rolland F, Moore B, Sheen J (2002) Sugar sensing and signaling in plants. Plant Cell 14:185–205

Romero-Puertas MC, Palma JM, Gómez M, Del Río LA, Sandalio LM (2002) Cadmium causes the oxidative modification of proteins in pea plants. Plant Cell Environ 25:677–686

Romero-Puertas MC, Corpas FJ, Rodríguez-Serrano M, Gómez M, Del Río LA, Sandalio LM (2007) Differential expression and regulation of antioxidative enzymes by cadmium in pea plants. J Plant Physiol 164:1346–1357

Rossman TG (2000) Cloning genes whose levels of expression are altered by metals: implications for human health research. Am J Ind Med 38:335–339

Rouhier N, Jacquot J-P (2002) Plant peroxiredoxins: alternative hydroperoxide scavenging enzymes. Photosynth Res 74:259–268

Rouhier N, Gelhaye E, Sautiere PE, Brun A, Laurent P, Tagu D, Gerard J, de Faÿ E, Meyer Y, Jacquot JP (2001) Isolation and characterization of a new peroxiredoxin from poplar sieve tubes that uses either glutaredoxin or thioredoxin as a proton donor. Plant Physiol 127:1299–1309

Rouhier N, Villarejo A, Srivastava M, Gelhaye E, Keech O, Droux M, Finkemeier I, Samuelsson G, Dietz KJ, Jacquot J-P et al (2005) Identification of plant glutaredoxin targets. Antioxid Redox Signal 7:919–929

Rouhier N, Couturier J, Jacquot J-P (2006) Genome-wide analysis of plant glutaredoxin systems. J Exp Bot 57:1685–1696

Ruan X, Luo F, Li D, Zhang J, Liu Z, Xu W, Huang G, Li X (2011) Cotton BCP genes encoding putative blue copper-binding proteins are functionally expressed in fiber development and involved in response to high-salinity and heavy metal stresses. Physiol Plant 141:71–83

Sagi M, Fluhr R (2006) Production of reactive oxygen species by plant NADPH Oxidases. Plant Physiol 141:336–340

Sahi SV, Sharma NC (2005) Phytoremediation of lead. In: Shtangeeva I (ed) Trace and ultratrace elements in plants and soils, Series advances in ecological researches. Witpress, Southampton, Boston, pp 209–222

Saifullah, Meers E, Qadir M, de Caritat P, Tack FMG, Du Laing G, Zia MH (2009) EDTA-assisted Pb phytoextraction. Chemosphere 74:1279–1291

Sarma H, Deka S, Deka H, Saikia RR (2011) Accumulation of heavy metals in selected medicinal plants. Rev Environ Contam Toxicol 214:63–86

Schreck E, Foucault Y, Geret F, Pradere P, Dumat C (2011) Influence of soil ageing on bioavailability and ecotoxicity of lead carried by process waste metallic ultrafine particles. Chemosphere 85:1555–1562

Schreck E, Foucault Y, Sarret G, Sobanska S, Cécillon L, Castrec-Rouelle M, Uzu G, Dumat C (2012) Metal and metalloid foliar uptake by various plant species exposed to atmospheric industrial fallout: mechanisms involved for lead. Sci Total Environ 427–428:253–262

Semane B, Cuypers A, Smeets K, Van Belleghem F, Horemans N, Schat H, Vangronsveld J (2007) Cadmium responses in *Arabidopsis thaliana*: glutathione metabolism and antioxidative defence system. Physiol Plant 129:519–528

Semane B, Dupae J, Cuypers A, Noben J-P, Tuomainen M, Tervahauta A, Kärenlampi S, Van Belleghem F, Smeets K, Vangronsveld J (2010) Leaf proteome responses of *Arabidopsis thaliana* exposed to mild cadmium stress. J Plant Physiol 167:247–254

Seregin IV, Shpigun LK, Ivanov VB (2004) Distribution and toxic effects of cadmium and lead on maize roots. Russ J Plant Physiol 51:525–533

Seth CS (2012) A review on mechanisms of plant tolerance and role of transgenic plants in environmental clean-up. Bot Rev. doi:10.1007/s12229-011-9092-x

Shah K, Kumar RG, Verma S, Dubey R (2001) Effect of cadmium on lipid peroxidation, superoxide anion generation and activities of antioxidant enzymes in growing rice seedlings. Plant Sci 161:1135–1144

Shahid M (2010) Lead-induced toxicity to Vicia faba L. in relation with metal cell uptake and speciation. Ph.D. Thesis. University of Toulouse, Toulouse, France

Shahid M, Pinelli E, Pourrut B, Silvestre J, Dumat C (2011) Lead-induced genotoxicity to *Vicia faba* L. roots in relation with metal cell uptake and initial speciation. Ecotoxicol Environ Saf 74:78–84

Shahid M, Arshad M, Kaemmerer M, Pinelli E, Probst A, Baque D, Pradere P, Dumat C (2012a) Long term field metal extraction by pelargonium: phytoextraction efficiency in relation with plant maturity. Int J Phytoremediation 14:493–505

Shahid M, Pinelli E, Dumat C (2012b) Review of Pb availability and toxicity to plants in relation with metal speciation; role of synthetic and natural organic ligands. J Hazard Mater 219–220:1–12

Shahid M, Dumat C, Silvestre J, Pinelli E (2012c) Effect of fulvic acids on lead-induced oxidative stress to metal sensitive *Vicia faba* L. Plant. Biol Fertil Soils 48:689–697

Shahid M, Dumat C, Aslam M, Pinelli E (2012d) Assessment of lead speciation by organic ligands using speciation models. Chem Spec Bioavailab 24:248–252

Shahid M, Xiong T, Castrec-Rouelle T, Leveque T, Dumat C (2013a) Water extraction kinetics of metals, arsenic and dissolved organic carbon from industrial contaminated poplar leaves. J Environ Sci 25:2451–9. doi:10.1016/S1001-0742(12)60197-1

Shahid M, Ferrand E, Schreck E, Dumat C (2013b) Behavior and impact of zirconium in the soil-plant system: plant update and phytotoxicity. Rev Environ Contam Toxicol 221:107–127

Shahid M, Xiong T, Masood N, Leveque T, Quenea K, Austruy A, Foucault Y, Dumat C (2013c) Influence of plant species and phosphorus amendments on metal speciation and bioavailability in a smelter impacted soil: a case study of food-chain contamination. J Soils Sediments. doi:10.1007/s11368-013-0745-8

Shahid M, Dumat C, Pourrut B, Silvestre J, Laplanche C, Pinelli E (2013d) Influence of EDTA and citric acid on lead-induced oxidative stress to *Vicia faba* roots. J Soils Sediments. doi:10.1007/s11368-013-0724-0

Shahid M, Austruy A, Echevarria G, Arshad M, Sanaullah M, Aslam M, Nadeem M, Nasim W, Dumat C (2014) EDTA-enhanced phytoremediation of heavy metals: a review. Soil Sediment Contam Int J 23:389–416. doi:10.1080/15320383.2014.831029

Sharma SS, Dietz K-J (2006) The significance of amino acids and amino acid-derived molecules in plant responses and adaptation to heavy metal stress. J Exp Bot 57:711–726

Sharma P, Dubey RS (2005) Lead toxicity in plants. Braz J Plant Physiol 17:35–52

Sharma SK, Goloubinoff P, Christen P (2008) Heavy metal ions are potent inhibitors of protein folding. Biochem Biophys Res Commun 372:341–345

Shen Y, Zhang Y, Chen J, Lin H, Zhao M, Peng H, Liu L, Yuan G, Zhang S, Zhang Z, Pan G (2013) Genome expression profile analysis reveals important transcripts in maize roots responding to the stress of heavy metal Pb. Physiol Plant 147:270–82. doi:10.1111/j.1399-3054.2012.01670.x

Sheng Z, Chaohai W, Chaodeng L, Haizhen W (2008) Damage to DNA of effective microorganisms by heavy metals: impact on wastewater treatment. J Environ Sci 20:1514–1518

Shi Q, Zhu Z (2008) Effects of exogenous salicylic acid on manganese toxicity, element contents and antioxidative system in cucumber. Environ Exp Bot 63:317–326

Shin L-J, Huang H-E, Chang H, Lin Y-H, Feng T-Y, Ger M-J (2011) Ectopic ferredoxin I protein promotes root hair growth through induction of reactive oxygen species in *Arabidopsis thaliana*. J Plant Physiol 168:434–440

Shulaev V, Cortes D, Miller G, Mittler R (2008) Metabolomics for plant stress response. Physiol Plant 132:199–208

Sies H (1993) Strategies of antioxidant defense. Eur J Biochem 215:213–219

Singh HP, Batish DR, Kaur G, Arora K, Kohli RK (2008) Nitric oxide (as sodium nitroprusside) supplementation ameliorates Cd toxicity in hydroponically grown wheat roots. Environ Exp Bot 63:158–167

Singh HP, Kaur S, Batish DR, Sharma VP, Sharma N, Kohli RK (2009) Nitric oxide alleviates arsenic toxicity by reducing oxidative damage in the roots of *Oryza sativa* (rice). Nitric Oxide 20:289–297

Singh R, Tripathi RD, Dwivedi S, Kumar A, Trivedi PK, Chakrabarty D (2010) Lead bioaccumulation potential of an aquatic macrophyte *Najas indica* are related to antioxidant system. Bioresour Technol 101:3025–3032

Singh NK, Rai UN, Tewari A, Singh M (2010) Metal accumulation and growth response in *Vigna radiata* L. inoculated with chromate tolerant rhizobacteria and grown on tannery sludge amended soil. Bull Environ Contam Toxicol 84:118–124

Singla-Pareek SL, Yadav SK, Pareek A, Mk R, Sopory SK (2006) Transgenic tobacco overexpressing glyoxalase pathway enzymes grow and set viable seeds in zinc-spiked soils. Plant Physiol 140:613–623

Šírová J, Sedlářová M, Piterková J, Luhová L, Petřivalský M (2011) The role of nitric oxide in the germination of plant seeds and pollen. Plant Sci 181:560–572

Smeets K, Ruytinx J, Semane B, Van Belleghem F, Remans T, Van Sanden S, Vangronsveld J, Cuypers A (2008) Cadmium-induced transcriptional and enzymatic alterations related to oxidative stress. Environ Exp Bot 63:1–8

Ströher E, Dietz K-J (2006) Concepts and approaches towards understanding the cellular redox proteome. Plant Biol (Stuttg) 8:407–418

Sun L-N, Zhang Y-F, He L-Y, Chen Z-J, Wang Q-Y, Qian M, Sheng X-F (2010) Genetic diversity and characterization of heavy metal-resistant-endophytic bacteria from two copper-tolerant plant species on copper mine wasteland. Bioresour Technol 101:501–509

Swanson S, Gilroy S (2010) ROS in plant development. Physiol Plant 138:384–392

Swanson SJ, Choi W-G, Chanoca A, Gilroy S (2011) In vivo imaging of Ca2+, pH, and reactive oxygen species using fluorescent probes in plants. Annu Rev Plant Biol 62:273–297

Szőllősi R, Varga IS, Erdei L, Mihalik E (2009) Cadmium-induced oxidative stress and antioxidative mechanisms in germinating Indian mustard (*Brassica juncea* L.) seeds. Ecotoxicol Environ Saf 72:1337–1342

Tak HI, Ahmad F, Babalola OO (2013) Advances in the application of plant growth-promoting rhizobacteria in phytoremediation of heavy metals. Rev Environ Contam Toxicol 223:33–52

Tan Y-F, O'Toole N, Taylor NL, Millar AH (2010) Divalent metal ions in plant mitochondria and their role in interactions with proteins and oxidative stress-induced damage to respiratory function. Plant Physiol 152:747–761

Tang K, Zhan J-C, Yang H-R, Huang W-D (2010) Changes of resveratrol and antioxidant enzymes during UV-induced plant defense response in peanut seedlings. J Plant Physiol 167:95–102

Tarrago L, Laugier E, Zaffagnini M, Marchand C, Le Maréchal P, Rouhier N, Lemaire SD, Rey P (2009) Regeneration mechanisms of Arabidopsis thaliana methionine sulfoxide reductases B by glutaredoxins and thioredoxins. J Biol Chem 284:18963–18971

Triantaphylidès C, Havaux M (2009) Singlet oxygen in plants: production, detoxification and signaling. Trends Plant Sci 14:219–228

Triantaphylidès C, Krischke M, Hoeberichts FA, Ksas B, Gresser G, Havaux M, Van Breusegem F, Mueller MJ (2008) Singlet oxygen is the major reactive oxygen species involved in photooxidative damage to plants. Plant Physiol 148:960–968

Trotter EW, Grant CM (2003) Non-reciprocal regulation of the redox state of the glutathione-glutaredoxin and thioredoxin systems. EMBO Rep 4:184–188

Turchi A, Tamantini I, Camussi AM, Racchi ML (2012) Expression of a metallothionein A1 gene of *Pisum sativum* in white poplar enhances tolerance and accumulation of zinc and copper. Plant Sci 183:50–56

Turton HE, Dawes IW, Grant CM (1997) *Saccharomyces cerevisiae* exhibits a yAP-1-mediated adaptive response to malondialdehyde. J Bacteriol 179:1096–1101

Tuteja N, Singh MB, Misra MK, Bhalla PL, Tuteja R (2001) Molecular mechanisms of DNA damage and repair: progress in plants. Crit Rev Biochem Mol Biol 36:337–397

Tuteja N, Ahmad P, Panda BB, Tuteja R (2009) Genotoxic stress in plants: shedding light on DNA damage, repair and DNA repair helicases. Mutat Res 681:134–149

USGS (United States Geological Survey) (2012) Assessed April 23, 2012. http://minerals.usgs.gov/minerals/pubs/commodity/zirconium/

Uzu G, Sobanska S, Aliouane Y, Pradere P, Dumat C (2009) Study of lead phytoavailability for atmospheric industrial micronic and sub-micronic particles in relation with lead speciation. Environ Pollut 157:1178–1185

Uzu G, Sobanska S, Sarret G, Muñoz M, Dumat C (2010) Foliar lead uptake by lettuce exposed to atmospheric fallouts. Environ Sci Technol 44:1036–1042

Uzu G, Sauvain J-J, Baeza-Squiban A, Riediker M, Sánchez Sandoval Hohl M, Val S, Tack K, Denys S, Pradère P, Dumat C (2011a) In vitro assessment of the pulmonary toxicity and gastric availability of lead-rich particles from a lead recycling plant. Environ Sci Technol 45:7888–7895

Uzu G, Sobanska S, Sarret G, Sauvain JJ, Pradère P, Dumat C (2011b) Characterization of lead-recycling facility emissions at various workplaces: major insights for sanitary risks assessment. J Hazard Mater 186:1018–1027

Vadas TM, Ahner BA (2009) Cysteine- and glutathione-mediated uptake of lead and cadmium into *Zea mays* and *Brassica napus* roots. Environ Pollut 157:2558–2563

Valko M, Rhodes CJ, Moncol J, Izakovic M, Mazur M (2006) Free radicals, metals and antioxidants in oxidative stress-induced cancer. Chem Biol Interact 160:1–40

Vanhoudt N, Vandenhove H, Horemans N, Wannijn J, Bujanic A, Vangronsveld J, Cuypers A (2010a) Study of oxidative stress related responses induced in *Arabidopsis thaliana* following mixed exposure to uranium and cadmium. Plant Physiol Biochem 48:879–886

Vanhoudt N, Vandenhove H, Horemans N, Wannijn J, Van Hees M, Vangronsveld J, Cuypers A (2010b) The combined effect of uranium and gamma radiation on biological responses and oxidative stress induced in *Arabidopsis thaliana*. J Environ Radioact 101:923–930

Vanhoudt N, Vandenhove H, Horemans N, Remans T, Opdenakker K, Smeets K, Bello DM, Wannijn J, Van Hees M, Vangronsveld J et al (2011) Unraveling uranium induced oxidative stress related responses in Arabidopsis thaliana seedlings. Part I: responses in the roots. J Environ Radioact 102:630–637

Verbruggen N, Hermans C, Schat H (2009) Molecular mechanisms of metal hyperaccumulation in plants. New Phytol 181:759–776

Verdoucq L, Vignols F, Jacquot JP, Chartier Y, Meyer Y (1999) In vivo characterization of a thioredoxin h target protein defines a new peroxiredoxin family. J Biol Chem 274:19714–19722

Verma S, Dubey RS (2003) Lead toxicity induces lipid peroxidation and alters the activities of antioxidant enzymes in growing rice plants. Plant Sci 164:645–655

Vuai SAH, Tokuyama A (2011) Trend of trace metals in precipitation around Okinawa Island, Japan. Atmos Res 99:80–84

Wallis JG, Browse J (2002) Mutants of arabidopsis reveal many roles for membrane lipids. Prog Lipid Res 41:254–278

Wahsha M, Bini C, Fontana S, Wahsha A, Zilioli D (2012) Toxicity assessment of contaminated soils from a mining area in Northeast Italy by using lipid peroxidation assay. J Geochem Explor 113:112–117

Wang L, Yang L, Yang F, Li X, Song Y, Wang X, Hu X (2010) Involvements of H_2O_2 and metallothionein in NO-mediated tomato tolerance to copper toxicity. J Plant Physiol 167:1298–1306

Wei ZH, Bai L, Deng Z, Zhong JJ (2011) Enhanced production of validamycin A by H_2O_2-induced reactive oxygen species in fermentation of *Streptomyces hygroscopicus* 5008. Bioresour Technol 102:1783–1787

Weyemi U, Dupuy C (2012) The emerging role of ROS-generating NADPH oxidase NOX4 in DNA-damage responses. Mutat Res 751:77–81. doi:10.1016/j.mrrev.2012.04.002

Whitaker C, Beckett RP, Minibayeva FV, Kranner I (2010) Production of reactive oxygen species in excised, desiccated and cryopreserved explants of *Trichilia dregeana* Sond. S Afr J Bot 76:112–118

Whiteside JR, Box CL, McMillan TJ, Allinson SL (2010) Cadmium and copper inhibit both DNA repair activities of polynucleotide kinase. DNA Repair (Amst) 9:83–89

Witkiewicz-Kucharczyk A, Bal W (2006) Damage of zinc fingers in DNA repair proteins, a novel molecular mechanism in carcinogenesis. Toxicol Lett 162:29–42

Wojas S, Clemens S, Skodowska A, Antosiewicz DM (2010) Arsenic response of AtPCS1- and CePCS-expressing plants—effects of external As (V) concentration on As-accumulation pattern and NPT metabolism. J Plant Physiol 167:169–175

Wonisch W, Hayn M, Schaur RJ, Tatzber F, Kranner I, Grill D, Winkler R, Bilinski T, Kohlwein SD, Esterbauer H (1997) Increased stress parameter synthesis in the yeast *Saccharomyces cerevisiae* after treatment with 4-hydroxy-2-nonenal. FEBS Lett 405:11–15

Xiang C, Werner BL, Christensen EM, Oliver DJ (2001) The biological functions of glutathione revisited in arabidopsis transgenic plants with altered glutathione levels. Plant Physiol 126:564–574

Xiao S, Chye M-L (2011) New roles for acyl-CoA-binding proteins (ACBPs) in plant development, stress responses and lipid metabolism. Prog Lipid Res 50:141–151

Xing S, Lauri A, Zachgo S (2006) Redox regulation and flower development: a novel function for glutaredoxins. Plant Biol (Stuttg) 8:547–555

Xu W, Li W, He J, Balwant S, Xiong Z (2009) Effects of insoluble Zn, Cd, and EDTA on the growth, activities of antioxidant enzymes and uptake of Zn and Cd in *Vetiveria zizanioides*. J Environ Sci (China) 21:186–192

Xu W, Li Y, He J, Ma Q, Zhang X, Chen G, Wang H, Zhang H (2010a) Cd uptake in rice cultivars treated with organic acids and EDTA. J Environ Sci (China) 22:441–447

Xu QS, Hu JZ, Xie KB, Yang HY, Du KH, Shi GX (2010b) Accumulation and acute toxicity of silver in *Potamogeton crispus* L. J Hazard Mater 173:186–193

Yadav SK (2010) Heavy metals toxicity in plants: an overview on the role of glutathione and phytochelatins in heavy metal stress tolerance of plants. S Afr J Bot 76:167–179

Yamauchi Y, Sugimoto Y (2010) Effect of protein modification by malondialdehyde on the interaction between the oxygen-evolving complex 33 kDa protein and photosystem II core proteins. Planta 231:1077–1088

Yan DYS, Lo IMC (2011) Enhanced multi-metal extraction with EDDS of deficient and excess dosages under the influence of dissolved and soil organic matter. Environ Pollut 159:78–83

Yang JL, Wang LC, Chang CY, Liu TY (1999) Singlet oxygen is the major species participating in the induction of DNA strand breakage and 8-hydroxydeoxyguanosine adduct by lead acetate. Environ Mol Mutagen 33:194–201

Yang Y, Wei X, Lu J, You J, Wang W, Shi R (2010) Lead-induced phytotoxicity mechanism involved in seed germination and seedling growth of wheat (*Triticum aestivum* L.). Ecotoxicol Environ Saf 73:1982–1987

Yeh C, Hung W, Huang H (2003) Copper treatment activates mitogen-activated protein kinase signalling in rice. Physiol Plant 119:392–399

Yeh C-M, Chien P-S, Huang H-J (2007) Distinct signalling pathways for induction of MAP kinase activities by cadmium and copper in rice roots. J Exp Bot 58:659–671

Yılmaz DD, Parlak KU (2011) Changes in proline accumulation and antioxidative enzyme activities in *Groenlandia densa* under cadmium stress. Ecol Indic 11:417–423

Yu S, Qin W, Zhuang G, Zhang X, Chen G, Liu W (2009) Monitoring oxidative stress and DNA damage induced by heavy metals in yeast expressing a redox-sensitive green fluorescent protein. Curr Microbiol 58:504–510

Zadák Z, Hyspler R, Tichá A, Hronek M, Fikrová P, Rathouská J, Hrnciaríková D, Stetina R (2009) Antioxidants and vitamins in clinical conditions. Physiol Res 58(Suppl 1):S13–S17

Zaffagnini M, Michelet L, Marchand C, Sparla F, Decottignies P, Le Maréchal P, Miginiac-Maslow M, Noctor G, Trost P, Lemaire SD (2007) The thioredoxin-independent isoform of chloroplastic glyceraldehyde-3-phosphate dehydrogenase is selectively regulated by glutathionylation. FEBS J 274:212–226

Zaffagnini M, Bedhomme M, Groni H, Marchand CH, Puppo C, Gontero B, Cassier-Chauvat C, Decottignies P, Lemaire SD (2012a) Glutathionylation in the photosynthetic model organism *Chlamydomonas reinhardtii*: a proteomic survey. Mol Cell Proteomics 11:M111.014142

Zaffagnini M, Bedhomme M, Lemaire SD, Trost P (2012b) The emerging roles of protein glutathionylation in chloroplasts. Plant Sci 185–186:86–96

Zawoznik GM, Tomaro M, Benavides M (2007) Endogenous salicylic acid potentiates cadmium-induced oxidative stress in *Arabidopsis thaliana*. Plant Sci 173:190–197

Zhang S, Klessig DF (2001) MAPK cascades in plant defense signaling. Trends Plant Sci 6:520–527

Zhang F, Zhang H, Wang G, Xu L, Shen Z (2009) Cadmium-induced accumulation of hydrogen peroxide in the leaf apoplast of *Phaseolus aureus* and *Vicia sativa* and the roles of different antioxidant enzymes. J Hazard Mater 168:76–84

Zhang X, Zhang S, Xu X, Li T, Gong G, Jia Y, Li L, Deng L (2010) Tolerance and accumulation characteristics of cadmium in *Amaranthus hybridus* L. J Hazard Mater 180:303–308

Zhao H, Xia B, Fan C, Zhao P, Shen S (2012) Human health risk from soil heavy metal contamination under different land uses near Dabaoshan Mine, Southern China. Sci Total Environ 417–418:45–54

Zhu C, Ding Y, Liu H (2011) MiR 398 and plant stress responses. Physiol Plant 143:1–9

Biological Responses of Agricultural Soils to Fly-Ash Amendment

Rajeev Pratap Singh, Bhavisha Sharma, Abhijit Sarkar, Chandan Sengupta, Pooja Singh, and Mahamad Hakimi Ibrahim

Contents

1 Introduction 46
2 Physico-Chemical Properties of Fly Ash (FA) 47
3 Biological Responses of Agricultural Soil to FA Amendment 49
    3.1 Physico-Chemical Responses of Soil to FA Amendment 49
    3.2 FA Management and the Soil Biochemical Cycle 52
    3.3 FA Management and Soil Microbial Dynamics 53
    3.4 Other Responses of Soil Health to Fly-Ash Amendment 54
4 Conclusions 55
5 Summary 55
References 56

R.P. Singh (✉) • B. Sharma
Institute of Environment and Sustainable Development, Banaras Hindu University, Varanasi, India
e-mail: rajeevprataps@gmail.com

A. Sarkar
Department of Botany, University of Kalyani, Kalyani, West Bengal, India

Research Laboratory for Biotechnology and Biochemistry (RLABB),
P.O. Box 13265, Kathmandu, Nepal

C. Sengupta
Department of Botany, University of Kalyani, Kalyani, West Bengal, India

P. Singh
Division of Bio-Resource, Paper and Coatings Technology,
School of Industrial Technology, Universiti Sains Malaysia, 11800 Penang, Malaysia

M.H. Ibrahim
Environmental Technology Division, School of Industrial Technology, Universiti Sains Malaysia, 11800 Penang, Malaysia

1 Introduction

Increased urbanization and industrialization worldwide has resulted in increased releases of solid waste, and enhanced environmental pollution around the globe. There are several categories of solid waste and these include sewage sludge, and municipal solid wastes (Singh et al. 2011). Fly Ash (FA), a coal combustion residue (CCR), is a major type of solid waste. The global dependence on coal as a major source of energy production, especially to produce electricity, has made FA a prime solid waste problem and a growing environmental pollutant. Proven global coal reserves have been estimated at 847 billion tons for the year 2007 (Sarkar et al. 2012). The USA has the largest share of global coal reserves (25.4%), followed by Russia (15.9%), China (11.6%) and India (8.6%) (Sarkar et al. 2012). Since India became independent in 1947, there has been a rapid increase in power generation, largely dominated by coal-based thermal generation constituting about 79% of total production. Energy production has increased from a capacity of 1,362 MW in 1947 to 120,000 MW in 2005. The Indian government plans to increase installed capacity to 300,000 MW by 2017 (Kumar et al. 2005; Vaidya 2009). India, like the United States, Russia and China, possesses abundant coal reserves, and coal-fueled generation of electricity is the common national policy (Singh et al. 2012; Sarkar et al. 2012).

During the combustion of coal several residues are produced. These include FA, bottom ash, flue gas desulphurization waste, fluidized bed boiler waste and coal gasification ash. FA is a residue of coal combustion (CCRs) that enters the flue gas stream. The nature of the FA produced largely depends on the quality and ash content of the coal that is burned. Indian coal is generally of lower grade than imported coals, and thereby has higher ash content (40%; CEA 2011).

The annual production of FA has increased from about 1.0 million metric tons (MT) in 1947 to about 112 MT during 2005. According to estimates from the FA Utilization Programme (FAUP), FA production is likely to reach 225 MT annually by 2017 (Kumar et al. 2005) (Fig. 1). Disposal of such an enormous amount of FA is a massive problem, particularly if it must be deposited in areas that surround thermal power stations. The major portion of FA produced in India is disposed of in ash ponds and in landfills; a minor proportion (<15%) is used to manufacture bricks, ceramics and cements (Pandey et al. 2009). The utilization of FA (3% of the 40 MT produced in 1994), has increased to ~38% of total production (viz., 112 MT) during 2004–05; this proportion is far below the global utilization rate (Dhadse et al. 2008; Singh et al. 2010) (Fig. 1). In India, 49% of FA is utilized in the cement industry, whereas only about 1% is used in the agricultural sector (Singh et al. 2010).

In agriculture, FA is primarily utilized as a soil amendment to buffer the soil pH (Phung et al. 1978). Such amendment improves soil texture (Fail and Wochok 1977; Chang et al. 1977) and soil nutrient status (Rautaray et al. 2003). However, the majority of the FA that is produced remains in ash storage ponds, and these deposits pose risks of several adverse effects to the environment.

In the present review, our aim is to address how FA can be utilized in global agriculture, and to provide the consequences of this use on soil health. Our major focus is

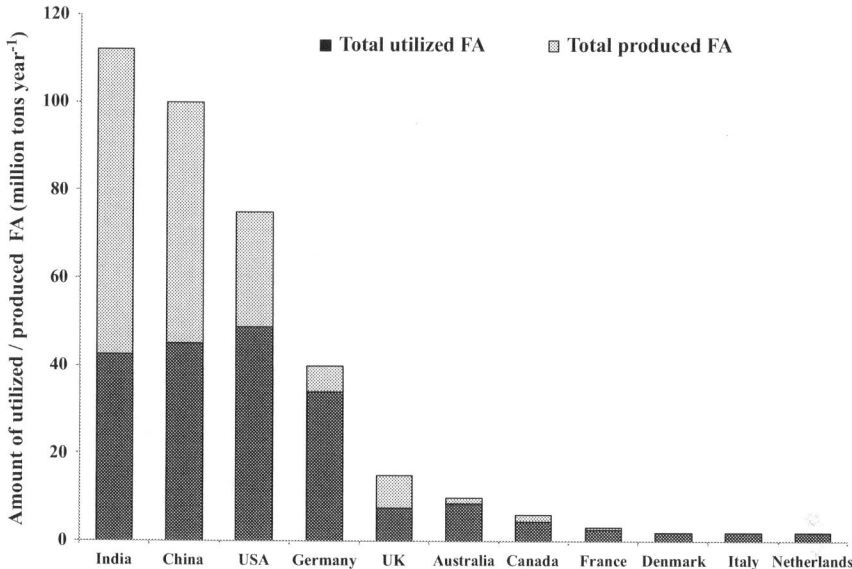

Fig. 1 The amount of FA produced and utilized in different countries. Source: Dhadse et al (2008)

to understand what the biological responses (i.e., physico-chemical, microbial, biochemical, etc.) are to FA-amended agricultural soils, and what effect FA amendment has on agricultural productivity. It is our intent to make this review useful for students and established researchers who work in the areas of soil nutritional dynamics and solid waste amendment. This review should also benefit some policy makers, who face the task of designing better and more sustainable approaches for managing solid waste pollution.

2 Physico-Chemical Properties of Fly Ash (FA)

The physico-chemical properties of FA primarily depend on the nature of the parent coal composition from which it comes, and secondly on the conditions under which the coal is combusted (Karapanagioti and Atalay 2001; Pandey and Singh 2010). Coal is a complex polymeric solid lacking any repeating monomeric units. FA is formed from the mineral matter in coal, and comprises a fine powder consisting of the non-combustible matter in coal, along with a small quantity of carbon that remains from incomplete combustion. FA is the finest of coal ash particles.

Physically, FA is comprised of very fine glass-like particles that are 0.01–100 mm in size (Davison et al. 1974; Jala and Goyal 2006). These FA particles have specific gravities of 2.1–2.6 g m^{-3} (Bern 1976), low to medium bulk density, a large surface area and very light texture. The specific chemical composition of FA depends on the quality of and conditions under which the parent coal was combusted (Jala and

Goyal 2006; Basu et al. 2009; Gupta et al. 2012). Some particles of FA are empty spheres (cenospheres), while others (plerospheres) are filled with small amorphous particles (Hodgson and Holliday 1966). FA constitutes a varied combination of amorphous and crystalline phases (usually considered as ferroaluminosilicate) (Lim and Choi 2014) and has a matrix similar to soil. It also contains about 69% of a fine-earthed fraction (i.e., clay silt) that derives from coal. Hodgson and Townsend (1973) reported that samples of fly-ash-particle fractions contained from 45 to 70% silt and 1 to 4% clay. The bulk density of different fly ashes varies from 1 to 1.8 g cm^3, whereas the pH ranges from 4.5 to 12.0, and depends on the S content of the parent coal (Plank and Martens 1974).

Alkalinity is an important FA characteristic, and results from the presence of Ca, Na, Mg and OH, along with certain other trace metals. Kunavanakrit (1993) reported that FA contained a high amount of Ca and Mg, both of which have high pH (11) and a high cation exchange capacity (CEC). The sub-bituminous and lignite coal ashes produce alkaline solutions when mixed with water. The degree of alkalinity depends on the Ca content, since this element is in the highly reactive CaO form, and is a major constituent of the fly-ash- forming $Ca(OH)_2$ (Hodgson et al. 1982). The characteristics of FA are greatly influenced by the particle size of its components. Particle size also affects the physical properties of fly-ash-amended soil.

Parameters that describe the chemical characteristics of coal include molecular weight, carbon aromaticity, normal aromatic and aliphatic structure and functional groups present. Coal quality is ranked by using several criteria: anthroxylon content, oxygen content, calorific value, ultimate analysis, fixed carbon content, etc. (Hodgson et al. 1982; Speight 2005). By and large, Indian coals have a high mineral matter %, low S content, high moisture, high ash content (Oliveira et al. 2014) and low calorific value (3,500–4,000 kcal kg^{-1}) (Gupta et al. 2012). The ash content of Indian coal varies between 15 and 30% and the S content is usually <1% (Srivastava 2003; Bhatt 2006). FA consists of approximately 95–99% of Si, Al, Fe and Ca oxides and about 0.5–3.5% of Na, P, K and S and the residual is trace elements.

Ahmaruzzaman (2010) described FA as mainly being composed of Si, Al, and Fe, with a major proportion of Ca, K, Na, Ti, along with other trace elements. Coal FA consists of SiO_2 (49–67%), Al_2O_3 (16–29%), Fe_2O_3 (4–10%), CaO (1–4%), MgO (0.2–2%), and SO_3 (0.1–2%) (Anon 2006; Singh et al. 2010). All metals present in soil are also found in fly ash. In Table 1, we compare the physico-chemical characteristics of FA and soil. The concentration of various elements that occur in FA varies with particle size (Khan and Khan 1996). A listing of elements present in FA includes the following: Si, Ca, Mg, Na, K, Cd, Pb, Cu, Co, Fe, Mn, Mo, Ni, Zn, B, F and Al (Tripathi et al. 2004, Gupta and Sinha 2000), and therefore, all important metals essential for plant growth and metabolism are present except organic C and N. The reason FA lacks any or much N is because it is volatilized from the coal (Singh and Yunus 2000). In contrast, FA has a high concentration of phosphorous (P) (400–8,000 mg P kg^{-1}). Unfortunately, this P is not readily available to plants, which may be due to its active interaction with Al, Fe and Ca present in alkaline FA (Gupta et al. 2012).

Table 1 A comparison of the physico-chemical properties of FA, an agricultural soil, and an FA-amended agricultural soil

Properties	Fly Ash (Tripathi et al. 2004)	Fly Ash (Gupta and Sinha 2008)	Soil (Tripathi et al. 2004)	FA amended soil (20% wt/wt) (Singh (2009) (PhD thesis, unpublished data))
pH	8.80	8.12	8.05	7.86
E. C. (mS cm^{-1})	7.61	3.54	0.23	3.477
Organic carbon (%)	1.17	1.7	43.40	0.537
Total nitrogen (%)	0.02	–	2.50	0.117
Total phosphorus (%)	0.14	–	1.06	–
Metals (mg kg^{-1})				
K	9,005.00	28,706.00	–	472.96
Na	5,200.00	41,321.00	–	396.74
Fe	4,150.00	20,054.00	2,850.00	1518.26
Zn	82.00	94.70	22.60	–
Cd	42.30	31.23	< 0.002	–
Pb	40.10	26.81	< 0.005	–
B	29.00	–	1.36	–
Ni	204.00	23.44	23.80	–

Several workers have reported the presence of radionuclides in fly ash; however, little information exists as to their impact (Gowiak and Pacynas 1980; Mittra et al. 2005; Papastefanou 2008). Mittra et al. (2005) analyzed the radioactivity (Bq kg^{-1}) of FA and recorded high radioactivity levels of ^{226}Ra, ^{228}Ac and ^{40}K in soil treated with FA at 40 t ha^{-1}. Moreover, Tadmore (1986) reported the radionuclides of uranium (U) and thorium (Th) series as components of fly ash.

FA is generally rich in toxic heavy metals (e.g., manganese, nickel, lead, etc.) and hazardous organic pollutants (e.g., polycyclic aromatic hydrocarbons, polychlorinated biphenyls, methyl sulphates, chlorinated dioxins and benzofurans (Wheatley and Sadhra 2004). Therefore, using FA in agriculture can result in higher accumulation of such toxic chemicals in food products, which, in turn, could pose human health issues.

3 Biological Responses of Agricultural Soil to FA Amendment

3.1 Physico-Chemical Responses of Soil to FA Amendment

The effect of amending soils with FA has been extensively investigated (Plank and Martens 1974; Elseewi and Page 1984; Jala and Goyal 2006). Kesh et al. (2003) reported FA as a repository of nutrients that assists in reclaiming alkaline and saline soils and improving soil properties. Amending soils with FA affects all soil physical

Table 2 The physico-chemical and biological responses of soil that has been amended with FA

Soil properties	Effect	References
Physical		
pH	Decrease	Pathan et al. (2003), Sinha and Gupta (2005), Gupta and Sinha (2006)
	Increase	Wong and Wong (1990), Jala and Goyal (2006)
Aggregate stability	Increase	Jala and Goyal (2006), Basu et al. (2009), Singh et al. (2010)
Bulk density	Decrease	Page et al. (1979), Singh et al. (2012a), Basu et al. (2009), Gupta et al. (2012)
Water holding capacity	Increase	Campbell et al. (1983), Page et al. (1979), Chang et al. (1977), Jala and Goyal (2006), Basu et al. (2009), Pandey and Singh (2010)
Porosity	Decrease	Page et al. (1979), Pandey and Singh (2010), Gupta et al. (2012)
Chemical		
Toxic elements (Cd, Pb, Ni etc.)	Increase	Gupta and Sinha (2006), Singh et al. (2010), Pandey and Singh (2010)
Fe, Cu, Zn, Mn	Increase	Tripathi et al. (2004), Gupta and Sinha (2006, 2008)
Electrical conductance	Increase	Adriano et al. (1980), Eary et al. (1990)
	Decrease	Gupta and Sinha (2006), Pandey and Singh (2010), Gupta et al. (2012)
Cation exchange capacity (CEC)	Decrease	Sinha and Gupta (2005), Gupta and Sinha. (2006), Jala and Goyal (2006)
Organic carbon / organic matter	Decrease	Gupta and Sinha (2006), Singh et al. (2010), Gupta et al. (2012)
Biological		
Microbial activity	Decrease	Adriano et al. (1978), Wong and Wong (1986), Saffigna et al. (1989)
	Increase	Schutter and Fuhrmann (2001)
Leachablity		
Pesticides	Decrease	Konstantinou and Albanis (2000); Singh et al. (2012b, 2013a, b)
Heavy meals	Increase	Natusch and Wallace (1974)

and chemical characteristics such as texture, bulk density, pH, water-holding capacity, electrical conductance (EC) (Chang et al. 1977; Pathan et al. 2003; Singh et al. 2012a) and particle size distribution (Sharma 1989) (Table 2). A gradual increase in the rate of fly-ash amendment (0% 10% 25%, up to 100% v/v) in normal field soils increased water-holding capacity, EC, and pH (Gupta and Sinha 2006, 2009).

Chemical properties of soil are also affected by adding fly ashes, since they are rich in heavy metal content (Singh et al. 2010, 2012a, Gupta and Sinha 2006, 2009) (Table 2). Campbell et al. (1983) reported that adding FA to soil @ 10% (wt/wt) increased the water holding capacity of soil by 7.2 and 413.2 times for fine and coarse sands, respectively. The water holding capacity of sandy soils is improved from the fine textured nature of fly ash; FA amendment is also known to reduce compaction of clay soils (Sharma and Kalra 2006).

FA amendment also increases the amounts of soluble major and minor inorganic constituents of soil, resulting in a higher EC value (Adriano et al. 1980; Eary et al. 1990; Jala and Goyal 2006; Basu et al 2009; Pandey and Singh 2010) (Table 2). The fly ashes from India are primarily alkaline in nature; hence, applying them increases soil pH from the rapid release of Ca, Na, Al and OH^- (Wong and Wong 1990; Sinha and Gupta 2005) (Table 2).

In addition to containing heavy-metals, FA also retains trace elements that may contaminate soil (Basu et al. 2009; Singh et al. 2010). The majority of trace metals are released at a pH value of approximately 9 (Ahmaruzzaman 2010). Addition of a minute amount of FA to soils can significantly boost solution pH. As pH increases, there is a decrease in trace metal desorption from FA (Theis and Wirth 1977). Fly ash, because of its hydroxide and carbonate salt content, has the ability to neutralize soil acidity (Pathan et al. 2003). However, using excessive amounts of FA to neutralize soil acidity can result in excessive soil alkalinity, particularly with unweathered fly ashes (Sharma et al. 1989). In fact, some acidic fly ashes are deliberately used for reclaiming alkaline soils (Table 2).

Pandey et al. (2009) studied the influence of amending garden soils with fly ash, in which *Cajanus cajan* L. was planted. The amendment altered accumulation and translocation of hazardous metals into edible plant parts. *Cajanus cajan* L. Plants were grown in containers, in which the concentrations of FA had been altered (0% 25%, 50% and 100% wt/wt). Amendment with FA at ratios from 25 to 100% in this garden soil increased the pH, the particle density, porosity and water holding capacity in comparison to controls from 3.47% to 26.39%, 3.98% to 26.14%, 37.50% to 147.92% and 163.16% to 318.42%, respectively. This amendment also decreased bulk density from 8.94 to 48.89% in the amended soil as compared to non-amended soil (Pandey et al. 2009).

Singh et al. (2012a) reported a decrease in NH_4^+, NO_3^-, total N, organic carbon (OC), organic matter (OM), available P, and CEC after rice was transplanted to a soil that had been amended with FA (0–20%). Reduced NH_4^+ and NO_3^- content from different levels of FA amendment was also reported by Singh and Agrawal (2010). Lee et al. (2006) reported increased soil pH and increased availability of Si, P, among other mineralogical components, in a Korean paddy field soil that was amended with fly ash; they concluded that FA can be utilized for improving the nutritional balance in a paddy field soil (Lee et al. 2006).

Generally, the bulk density of soil declined with the addition of fly ash, which in turn reduced porosity and increased water holding capacity (Page et al. 1979; Pandey and Singh 2010). Several workers have reported that FA amendment significantly increases the water holding capacity of the amended soil. Although FA itself does not retain water efficiently, amending sandy and loamy soils with it increased water holding capacity by 8% (Chang et al. 1977). Singh and Agrawal (2010) reported a significant improvement in levels of soil nutrients (e.g., Na, K, Ca, Mg, and Fe) when increasing rates of FA were used to amend soils at Varanasi, India. The high boron (B) level in FA restricts its utilization in crop production (Aitken and Bell 1985). However, if the FA is properly weathered the problem with B can be overcome. FA has a liming effect on soils that increases calcium and hydroxide ion mobility, which in turn enriches bacterial growth (Surridge et al. 2009). However, high levels of toxic heavy metals that can be transferred to soils from adding FA (Page et al. 1979) can hamper normal microbial metabolic processes (Pandey and Singh 2010).

3.2 FA Management and the Soil Biochemical Cycle

Biological indicators are biological species that can be used to monitor environmental or ecosystem health. Biological indicators are often employed to represent some aspect of the living soil and its environment. Such indicators generally respond more rapidly to changes in the soil environment than do physical or chemical indicators (Anderson and Gray 1990; Pascual et al. 2000; Singh et al. 2011). Additionally, biological indicators are sensitive tools for detecting changes in soil conditions that may occur (Singh et al. 2011). Microbes are vital constituents of the soil environment that contribute to the degradation of organic matter and make nutrients more available to other soil organisms. The responses of microbes to the addition of FA have been explored in several studies that we will describe below, although there is a paucity of data for direct effects on the microbes themselves.

In the soil system, soil enzymes play a key biochemical role in organic matter decomposition (Burns 1983; Chròst 1991; Sinsabaugh et al. 1991). Enzymes are critical for catalyzing several reactions that are essential for life processes of soil micro-organisms; these include stabilizing the soil structure, nutrient cycling, decomposition of organic wastes and organic matter formation (Dick et al. 1994). These soil enzymes are continuously being synthesized, accumulated, inactivated and/or decomposed, and therefore play an important function in agriculture, mainly via assisting nutrient cycling (Tabatabai 1994; Dick 1997).

Each and every soil hosts a group of enzymes that perform metabolic processes (McLaren 1975), the presence and titers of which depend on the soil's physico-chemical, microbiological and biochemical properties. Because soil enzymes have such a critical role, they respond so quickly to changes in soil management practices and are easy to measure, knowing more about their function potentially helps in assessing the prevailing biological status and function of soils (Dick 1997; Bandick and Dick 1999). Soil enzymes often significantly affect soil biology, environmental management strategies, and growth and nutrient uptake of plants that inhabit ecosystems.

Soil fungi comprise at least 75–95% of soil microbial biomass, and along with bacteria contribute ~90% of the total energy flux to the organic matter decomposition in soil (Paul and Clark 1996). Soil enzyme activity is especially important for fertility. Soil enzymes are routinely measured to provide a biological index of soil fertility. This index serves as an indicator for several biological processes in soil. In general, the enzymatic activities of soil enzymes are used to reflect outcomes resulting from agricultural cultivation, and the existence of different soil properties, and pedological amendments (Skujins 1978; Ceccanti et al. 1993).

Adding FA to soil stimulates enzyme activity (viz., dehydrogenase, urease and phosphatases, etc.; Pati and Sahu 2004). As mentioned, amending soils with FA adds many elements (e.g., C, K, Ca, Mg, Cu, Zn and Mn), and these elements may alter the chemical and physico-chemical properties of the soils to which they are added (Yeledhalli et al. 2007).

The amount of microbial biomass present is commonly used to characterize the microbiological status of soils (Nannipieri et al. 1990), and to evaluate the effect of soil management practices (Perrott et al. 1992). Soil microbial biomass is a sound indicator of soil health, because such biomass regulates nutrient cycling and acts as a highly labile source of nutrients that are available to plants (Jenkinson and Ladd 1981). Rippon and Wood (1975) attributed increased microbial populations in a soil to the addition of FA . However, higher FA amendment levels sometimes resulted in deposition of excessive amounts of certain toxic elements (e.g., As and B) in soil, and such deposition negatively affected the normal soil microbial dynamics and activity (Lim and Choi 2014). FA amendment of soil may benefit fungi and gram-negative bacteria more than other components of the soil microbial community (Schutter and Fuhrmann 2001).

Soil microbial biomass and dehydrogenase activity were reported to be highest at a FA amendment rate of 10% (wt/wt), because at this rate reasonable levels of nutrients were provided to microorganisms for carrying out various metabolic activities (Wong and Wong 1986; Saffigna et al. 1989). Microbial activity declined when FA was added at levels of more than 10% (Wong and Wong 1986; Saffigna et al. 1989). This decline may have resulted from reduced substrate availability that was associated with accumulation of persistent lignite-derived organic carbon compounds (Rumpel et al. 1998). Gaind and Gaur (2004) reported that *Azotobacter chroococcum*, *Azospirillum brasilense* and *Bacillus circulans* showed their maximum viability when FA alone was applied to soil, whereas *Pseudomonas striata* proliferated most in soil-FA (1:1) applications. Generally, the effects of FA applications on soil aggregation, together with the effects of growing plants on soil microbial diversity may favor plant growth and soil revival. Wong and Wong (1987) found that the application of FA increased microbial respiration in a sandy soil and decreased it in a sandy loam soil. Arthur et al. (1984) concluded that lower rates of FA applied to soil had a modest impact on microbial activity, but higher rates inhibited microorganisms. Schutter and Fuhrmann (2001) reported that amending degraded subsoil with FA caused an increased density of the microbial community.

3.3   FA Management and Soil Microbial Dynamics

As for other major solid wastes, utilization of FA in agriculture has gained popularity worldwide in the past few decades (Singh and Agrawal 2008; Singh et al. 2012). More recently, researchers have studied the effects of FA on soil health, especially the effects on soil–microbial interactions and dynamics (Sarkar et al. 2012). Modern day '-omics' approaches represent state-of-the-art technologies that offer prospects for a major breakthrough in soil – microbial dynamics. The '-omics' have provided modern day researchers with better tools to identify and evaluate microbial diversity in soil, water and air under diverse environmental conditions (Schneider and Riedel 2010). Integrated genomics and proteomics approaches promise to be swift and effective systems for analyzing and deducing gene function in living organisms at genome (*genomics*), transcript (*transcriptomics*), and protein (*proteomics*) levels (Sarkar et al. 2012; Agrawal et al. 2013). These three approaches are commonly referred as the multi-parallel '-omics' approaches in modern biology (Sarkar et al. 2010; Zargar et al. 2011). Recently, researchers have started to work with 'genome' and 'proteome' samples that are directly isolated from environment (Sarkar and Agrawal 2012). These sample entities are termed the 'metagenome' and the 'metaproteome', respectively. The *in-vivo* and *in-vitro* '-omics' approaches have significantly contributed to the evaluation of soil – microbial dynamics in many ecosystems. By using a metagenomics approach Sanapareddy et al. 2009) generated 378,601 sequences by pyrosequencing (by using 454-FLX technology) of DNA samples collected from an activated sludge basin of a wastewater treatment plant in Charlotte, North Carolina, USA. These authors identified a significant number of microbial communities in the sludge basin that might be useful for improving soil health. Wang et al. 2011) employed a metaproteomics approach through in-depth two-dimensional gel electrophoresis (2DGE), coupled with matrix-assisted laser desorption/ionization time-of-flight mass spectrometer (MALDI-TOF/TOF-MS), and identified nearly 122 proteins, constituting a metaproteome of a plant-microbe complex that existed in a crop rhizospheric soil. Other researchers have also utilized '-omics', particularly metagenomics and metaproteomics approaches. Such techniques allow improved discernment of microbial dynamism in soil samples under diverse environmental conditions, and the contributions of microbes to soil health (Schneider and Riedel 2010).

3.4 Other Responses of Soil Health to Fly-Ash Amendment

FA affects aspects of soil health not described above (Ahmaruzzaman 2010) (Table 2). In particular, it is known that FA hinders the normal leaching pattern of metals in soil. The pH, and chemical composition of a soil, as well as the FA used to amend a soil are all important variables that can influence the leaching behaviour of heavy metals (Becker et al. 2013) (Table 2). Amending agricultural soils with FA is known to restrict the normal soil leaching pattern of pesticides, and to boost pesticide retention (Singh et al. 2012b, 2013a, b). Application of FA to soils at the 20–30% level has been reported to detoxify 2, 4-D, alachlor and metolachlor in soil (Albanis et al. 1992, 1998). Konstantinou and Albanis (2000) reported that amending soil with FA up to 25% can immobilize atrazine, propazine, prometryne, molinate, propachlor and propanil herbicides. Singh et al. (2013a, b) reported that FA amendment in soil did not show an adverse effect on weed control efficacy of the herbicides metribuzin and metsulfuron-methyl. Hence, it is conceivable that FA could be used to amend soils in ways to help manage herbicide runoff and leaching losses.

4 Conclusions

Our main conclusions from reviewing the cogent literature on fly ash amendment of agricultural soils and from preparing this review are as follows:

1. Fly ash is a waste product from coal combustion process, and is a potential resource for amending agricultural soils to provide several essential plants nutrients. However, organic C and N are not among these nutrients.
2. When amending agricultural soils with FA, the appropriate methods and amounts used will depend on soil type, nature of the cultivated crop, prevailing climatic conditions and the characteristics of the FA used.
3. FA has a very high affinity for organic pesticides. Therefore, using it as a soil amendment can boost pesticide retention in agricultural soils.
4. Although applying FA in normal agricultural practice may benefit plant nutrition, it has a downside of potentially enhancing contamination by heavy metals in ways that affect ground water, well (drinking) water, and food chain organisms.
5. Harmful effects may result from applying FA to amend agricultural soils. Harm may come from enhanced levels of natural radioactivity (from FA) and from increased levels of toxic heavy metals that could contaminate food or feed. Therefore, care must be taken when FA is to be used as an agricultural soil amendment.
6. FA amendment in agriculture is undoubtedly in its infancy, and requires further study, particularly on dose-response relationships, before it can quality for large scale application in global agriculture.

5 Summary

The volume of solid waste produced in the world is increasing annually, and disposing of such wastes is a growing problem. Fly ash (FA) is a form of solid waste that is derived from the combustion of coal. Research has shown that fly ash may be disposed of by using it to amend agricultural soils. This review addresses the feasibility of amending agricultural field soils with fly ash for the purpose of improving soil health and enhancing the production of agricultural crops. The current annual production of major coal combustion residues (CCRs) is estimated to be ~600 million t worldwide, of which about 500 million t (70–80%) is FA (Ahmaruzzaman 2010). More than 112 million t of FA is generated annually in India alone, and projections show that the production (including both FA and bottom ash) may exceed 170 million t per annum by 2015 (Pandey et al. 2009; Pandey and Singh 2010). Managing this industrial by-product is a big challenge, because more is produced each year, and disposal poses a growing environmental problem.

Studies on FA clearly shows that its application as an amendment to agricultural soils can significantly improve soil quality, and produce higher soil fertility. What FA application method is best and what level of application is appropriate for any one soil depends on the following factors: type of soil treated, crop grown, the prevailing agro climatic condition and the character of the FA used. Although utilizing FA in agricultural soils may help address solid waste disposal problems and may enhance agricultural production, its use has potential adverse effects also. In particular, using it in agriculture may enhance amounts of radionuclides and heavy metals that reach soils, and may therefore increase organism exposures in some instances.

Acknowledgement Authors, both BS and RPS, are thankful to UGC, New Delhi (P01/679) as well as Banaras Hindu University, Varanasi for necessary help. AS acknowledges the financial help in the form of DBT-RA from Department of Biotechnology, Government of India, India. AS is also thankful to the Dr. Ganesh Kumar Agrawal, RLABB, Nepal for allowing this collaborative work. CS acknowledges the financial help from DST - PURSE program to Department of Botany, University of Kalyani, from Department of Science and Technology, Govt. of India, India.RPS, PS and MHI, are thankful to University Sains Malaysia, Malaysia for necessary help.

References

Adriano DC, Woodford TA, Ciravolo TG (1978) Growth and elemental composition of corn and bean seedlings as influenced by soil application of coal ash. J Environ Qual 7:416–421

Adriano DC, Page AL, Elseewi AA, Chang AC, Straughan I (1980) Utilization and disposal of fly ash and other coal residues in terrestrial ecosystems: a review. J Environ Qual 9:333–344

Agrawal GK, Sarkar A, Righetti PG, Pedreschi R, Carpentier S et al (2013) A decade of plant proteomics and mass spectrometry: translation of technical advancements to food security and safety issues. Mass Spectrom Rev 32(5):335–365

Ahmaruzzaman M (2010) A review on the utilization of fly ash. Prog Energ Combust Sci 36(3):327–363

Aitken RL, Bell LC (1985) Plant uptake and phytotoxicity of boron in Australian fly ashes. Plant Soil 84:245–257

Albanis TA, Tzialla C, Pomonis PJ (1992) The influence of fly ash on 2, 4dichlorophenoxy-acetic acid persistence in corn cultivation and soil. Sci Total Environ 123:481–489

Albanis TA, Danis TG, Kourgia MG (1998) Adsorption-desorption studies of selected chlorophenols and herbicides and metal release in soil mixtures with fly ash. Environ Technol 19:25–34

Anderson TH, Gray TRG (1990) Soil microbial carbon uptake characteristics in relation to soil management. FEMS Microbiol Lett 74:11–19

Anon (2006) State of the environment, vol 3(1). Flyash Management, Orissa. Envis News Letter, Centre for Environmental Studies, Forest and Environment Department, Government of Orissa

Arthur MF, Zwick TC, Tolle DA, VanVoris P (1984) Effects of fly ash on microbial CO2 evolution from an agricultural soil. Water Air Soil Pollut 22:209–216

Bandick AK, Dick RP (1999) Field management effects on soil enzyme activities. Soil Biol Biochem 31:1471–1479

Basu M, Pande M, Bhadoria PBS, Mahapatra SC (2009) Potential fly-ash utilization in agriculture: a global review. Prog Nat Sci 19:1173–1186

Becker J, Aydilek AH, Davis AP, Seagren EA (2013) Evaluation of leaching protocols for testing of high-carbon coal fly ash–soil mixtures. J Environ Eng 139:642–653

Bern J (1976) Residues from power generation: processing, recycling and disposal, Land Application of Waste Materials, Soil Conservation. Society of American Ankeny, Iowa, pp 226–248

Bhatt MS (2006) Effect of ash in coal on the performance of coal fired thermal power plants. Part I: primary energy effects. Energ Source Part A 28:25–41

Burns RG (1983) Extracellular enzyme-substrate interactions in soil. In: Slater JH, Wittenbury R, Wimpenny JWT (eds) Microbes in their natural environment. Cambridge University Press, London, pp 249–298

Campbell DJ, Fox WE, Aitken RL, Bell LC (1983) Physical characteristics of sands amended with fly ash. Aust J Soil Res 21:147–154

CEA (2011) Operation performance of generating stations in the country during the year 2010–11—an overview. CEA, New Delhi

Ceccanti B, Pezzarossa B, Gallardo-Lancho FJ, Masciandaro G (1993) Bio-tests as markers of soil utilization and fertility. Geomicrobiol J 11:309–316

Chang AC, Lund LJ, Page AL, Warneke JE (1977) Physical properties of fly ash amended soils. J Environ Qual 6:267–270

Chròst RJ (1991) Environmental control of the synthesis and activity of aquatic microbial ectoenzymes. In: Chrost RJ (ed) Microbial enzymes in aquatic environments. Springer, New York, pp 29–59

Davison RL, Natusch DFS, Wallace JR, Evans CA Jr (1974) Trace elements in fly ash: dependence of concentration on particle size. Environ Sci Technol 8:1107–1113

Dhadse S, Pramilla K, Bhagia LJ (2008) Fly ash characterization, utilization and Government initiatives in India—a review. J Sci Ind Res India 67:11–18

Dick RP (1997) Soil enzyme activities as integrative indicators of soil health. In: Pankhurst CE, Doube BM, Gupta VVSR (eds) Biological indicators of soil health. CAB International, Wellingford, pp 121–156

Dick RP, Sandor JA, Eash NS (1994) Soil enzyme activities after 1500 years of terrace agriculture in the Colca Valley, Peru. Agric Ecosyst Environ 50:123–131

Eary LE, Rai D, Mattigod SV, Ainsworth CC (1990) Geochemical factors controlling the mobilization of inorganic constituents from fossil fuel combustion residues. II. Review of the minor elements. J Environ Qual 19:202–214

Elseewi AA, Page AL (1984) Molybdenum enrichment of plants grown on fly ash treated soils. J Environ Qual 13:394–398

Fail JL, Wochok ZS (1977) Soyabean growth on fly ash amended strip mine spoils. Plant Soil 48:473–484

Gaind S, Gaur AC (2004) Evaluation of fly ash as a carrier for diazotrophs and phosphobacteria. Bioresour Technol 95:187–190

Gowiak BJ, Pacyna JM (1980) Radiation dose due to atmospheric releases from coal-fired power stations. Int J Environ Stud 16:23–29

Gupta AK, Sinha S (2006) Role of *Brassica juncea* L. Czern. (var. vaibhav) in the phytoextraction of Ni from soil amended with fly-ash: selection of extractant for metal bioavailability. J Hazard Mater 136:371–378

Gupta AK, Sinha S (2008) Decontamination and/or revegetation of fly ash dykes through naturally growing plants. J Hazard Mater 153:1078–1087

Gupta AK, Sinha S (2009) Growth and metal accumulation response of *Vigna radiata* L. var PDM 54 (mung bean) grown on fly ash-amended soil: effect on dietary intake. Environ Geochem Health 31:463–473

Gupta AK, Singh RP, Ibrahim MH, Byeong-Kye (2012) Agricultural utilization of fly ash and its consequences. In: Eric Lichtfouse (ed) Sustainable agriculture reviews, vol 8. Springer, pp 269–286

Hodgson DR, Holliday R (1966) The agronomic properties of pulverized fuel ash. Chem Ind 20:785–790

Hodgson DR, Townsend WN (1973) The amelioration and revegetation of pulverized fuel ash. In: Hutnik RJ, Davis G (eds) Ecology and reclamation of devastated land, vol II. Gordon and Breach, New York, pp 247–271

Hodgson L, Dyer D, Brown DA (1982) Neutralization and dissolution of high calcium fly ash. J Environ Qual 11(1):93

Jala S, Goyal D (2006) Fly ash as a soil ameliorant for improving crop production—a review. Bioresour Technol 97:1136–1147

Jenkinson DS, Ladd JN (1981) Microbial biomass in soil: measurement and turnover. Soil Biol Biochem 5:415–417

Karapanagioti HK, Atalay AS (2001) Laboratory evaluation of ash materials as acid disturbed land amendments. Glob Nest 3(1):11–21

Kesh S, Kalra N, Sharma SK, Chaudhary A (2003) Fly ash incorporation effects on soil characteristics and growth and yield of wheat. Asia Pacific J Environ Dev 4:53–69

Khan MR, Khan MW (1996) The effect of fly-ash on plant growth and yield of tomato. Environ Pollut 92:105–111

Konstantinou IK, Albanis TA (2000) Adsorption–desorption studies of selected herbicides in soil-fly ash mixtures. J Agric Food Chem 48:4780–4790

Kumar V, Mathur M, Sinha SS (2005) A case study: manifold increase in fly ash utilization in India. Fly Ash Utilization Programme (FAUP), TIFAC, DST, New Delhi

Kunavanakrit W (1993) General properties of lignite fly ash. In: Conference on the potential of Lignite Fly Ash Utilization, EGAT, Thailand, pp 2–15

Lee H, Ha HS, Lee CS, Lee YB, Kim PJ (2006) Fly ash effect on improving soil properties and rice productivity in Korean paddy soil. Bioresour Technol 97:1490–1497

Lim SS, Choi WJ (2014) Changes in microbial biomass, CH4 and CO2 emissions, and soil carbon content by fly ash co-applied with organic inputs with contrasting substrate quality under changing water regimes. Soil Biol Biochem 68:494–502

McLaren AD (1975) Soil as a system of humus and clay immobilised enzymes. Chem Scripta 8:97–99

Mittra BN, Karmakar S, Swain DK, Ghosh BC (2005) Fly-ash a potential source of soil amendment and a component of integrated plant nutrient supply system. Fuel 84:1447–1451

Nannipieri P, Grego S, Ceccanti B (1990) Ecological significance of the biological activity in soil. Soil Biol Biochem 6:293–355

Natusch DFS, Wallace JR (1974) Urban aerosol toxicity: the influence of particle size. Science 186:695–699

Oliveira MLS, Marostega F, Taffarel SR, Saikia BK, Waanders FB, DaBoit K, Baruah BP, Silva LFO (2014) Nano-mineralogical investigation of coal and fly ashes from coal-based captive power plant (India): an introduction of occupational health hazards. Sci Total Environ 468–469:1128–1137

Page AL, Elseewi AA, Straughan IR (1979) Physical and chemical properties of fly ash from coal-fired power plants with special reference to environmental impacts. Residue Rev 71:83–120

Pandey VC, Singh N (2010) Impact of fly ash incorporation in soil systems. Agric Ecosyst Environ 136:16–27

Pandey VC, Abhilash PC, Upadhyay RN, Tewari DD (2009) Application of fly ash on the growth performance and translocation of toxic heavy metals within *Cajanus cajan* L.: implication for safe utilization of fly ash for agricultural production. J Hazard Mater 166:255–259

Papastefanou C (2008) Radioactivity of coals and fly ashes. J Radioanal Nucl Chem 275:29–35

Pascual JA, Garcia C, Hernandez T, Moreno JL, Ros M (2000) Soil microbial activity as a biomarker of degradation and remediation processes. Soil Biol Biochem 32:1877–1883

Pathan SM, Aylmore LAG, Colmer TD (2003) Soil properties and turf growth on a sandy soil amended with fly ash. Plant Soil 256:103–114

Pati SS, Sahu SK (2004) CO_2 evaluation and enzyme activities (dehydrogenase, protease and amylase) of fly ash amended soil in presence and absence of earthworms (Under laboratory condition). Geo Derma 118:289–301

Paul EA, Clark FE (1996) Soil microbiology and biochemistry. Academic, San Diego, CA, 340 p

Perrott KW, Sarathchandra SU, Dow BW (1992) Seasonal and fertilizer effects on the organic cycle and microbial biomass in a hill country soil under pasture. Aust J Soil Res 30:383–394

Phung HT, Lund LJ, Page AL (1978) Potential use of fly ash as a liming material. In: Adriano DC, Brisbin IL (eds) Environmental chemistry and cycling processes, CONF-760429. US Department of Commerce, Springfield, VA, pp 504–515

Plank CO, Martens DC (1974) Boron availability as influenced by application of fly ash to soil. Soil Sci Soc Am Proc 38:974–977

Rautaray SK, Ghosh BC, Mittra BN (2003) Effect of fly ash, organic wastes and chemical fertilizers on yield, nutrient uptake, heavy metal content and residual fertility in a rice-mustard cropping sequence under acid lateritic soils. Bioresour Technol 90:275–283

Rippon JE, Wood MJ (1975) Microbiological aspects of pulverized fuel ash. In: Chadwick MJ, Goodman GT (eds) The ecology of resource degradation and renewal. Wiley, New York, pp 331–349

Rumpel C, Knicker H, Kogel-Knaber I, Skjiemstad JO, Huuetti RF (1998) Types and chemical composition of organic matter in reforested lignite-rich mine soils. Geoderma 86:123–142

Saffigna PG, Powlson DS, Brookes PC, Thomas GA (1989) Influence of sorghum residues and tillage on soil organic matter and soil microbial biomass in an Australian vertisol. Soil Biol Biochem 21:759–765

Sanapareddy N, Hamp TJ, Gonzalez LC, Hilger HA, Fodor AA, Clinton SM (2009) Molecular diversity of a North Carolina wastewater treatment plant as revealed by pyrosequencing. Appl Environ Microb 75(6):1688–1696

Sarkar A, Agrawal SB (2012) Evaluating the response of two high yielding Indian rice cultivars against ambient and elevated levels of ozone by using open top chambers. J Environ Manage 95:S19–S24

Sarkar A, Rakwal R, Agrawal SB, Shibato J, Ogawa Y, Yoshida Y, Agrawal GK, Agrawal M (2010) Investigating the impact of elevated levels of ozone on tropical wheat using integrated phenotypical, physiological, biochemical and proteomics approaches. J Proteome Res 9(9):4565–4584

Sarkar A, Singh A, Agarawal SB (2012) Utilization of fly ash as soil amendments in agricultural fields on North-Eastern gangetic plains of India: potential benefits and risks assessments. Bull Nat Inst Ecol 23(1–2):9–20

Schneider T, Riedel K (2010) Environmental proteomics: analysis of structure and function of microbial communities. Proteomics 10:785–798

Schutter ME, Fuhrmann JJ (2001) Soil microbial community responses to fly ash amendment as revealed by analyses of whole soils and bacterial isolates. Soil Biol Biochem 33:1947–1958

Sharma S (1989) Fly ash dynamics in soil water systems. Crit Rev Environ Control 19:251–275

Sharma SK, Kalra N (2006) Effect of fly ash incorporation on soil properties and plant productivity—a review. J Sci Ind Res India 65:383–390

Sharma S, Fulekar MH, Jayalakshmi CP, Straub CP (1989) Fly ash dynamics in soil-water systems. Crit Rev Environ Control 19:251–275

Singh A, Agrawal SB (2010) Response of mung bean cultivars to fly ash: growth and yield. Ecotoxicol Environ Saf 73:1950–1958

Singh RP, Agrawal M (2008) Potential benefits and risks of land application of sewage sludge. Waste Manage 28:347–358

Singh N, Yunus M (2000) Environmental impacts of fly-ash. In: Iqbal M, Srivastava PS, Siddiqui TO (eds) Environmental hazards: plant and people. CBS, New Delhi, pp 60–79

Singh RP, Gupta AK, Ibrahim MH, Mittal AK (2010) Coal fly ash utilization in agriculture: its potential benefits and risks. Rev Environ Sci Biotechnol 9:345–358

Singh RP, Singh P, Ibrahim MH, Hashim R (2011) Land application of sewage sludge: physicochemical and microbial response. Rev Environ Contam Toxicol 214:41–61

Singh A, Sarkar A, Agrawal SB (2012a) Assessing the potential impact of fly ash amendments on Indian paddy field with special emphasis on growth, yield, and grain quality of three rice cultivars. Environ Monit Assess 184:4799–814

Singh N, Raunaq, Singh SB (2012b) Effect of fly ash on sorption behaviour of metribuzin in agricultural soils. J Environ Sci Health B47:89–98

Singh N, Raunaq, Singh SB (2013a) Reduced downward mobility of metribuzin in fly ash-amended soils. J Environ Sci Health B 48:587–59

Singh N, Singh SB, Raunaq, Das TK (2013b) Effect of fly ash on persistence, mobility and bio-efficacy of metribuzin and metsulfuron-methyl in crop fields. Ecotoxicol Environ Safety 97:236–241

Sinha S, Gupta AK (2005) Translocation of metals from fly ash amended soil in the plant of *Sesbania cannabina* L. Ritz: effect on antioxidants. Chemosphere 61:1204–1214

Sinsabaugh RL, Antibus RK, Linkins AE (1991) An enzymic approach to the analysis of microbial activity during plant litter decomposition. Agric Ecosyst Environ 34:43–54

Skujins J (1978) Soil enzymology and fertility index-a fallacy? History of abiotic soil enzyme research. In: Burns RG (ed) Soil enzymes. Academic, London, pp 1–49

Speight JG (2005) Handbook of coal analysis. Wiley Interscience, Hoboken, NJ

Srivastava SK (2003) Recovery of sulphur from very high ash fuel and fine distributed pyritic sulphur containing coal using ferric sulphate. Fuel Process Technol 84:37–46

Surridge AKJ, Merwe A, Kruger R, (2009) Preliminary microbial studies on the impact of plants and South African fly ash on amelioration of crude oil polluted soils. In World of Coal Ash (WOCA) conference, May 4–7

Tabatabai MA (1994) Soil enzymes. In: Weaver RW, Angle JS, Bottomley PS (eds) Methods of soil analysis, part 2. Microbiological and biochemical properties. SSSA Book Series No. 5. Soil Science Society of America, Madison, WI, pp 775–833

Tadmore J (1986) Radioactivity from coal-fired power plants: a review. J Environ Radioact 4:177–204

Theis TL, Wirth JL (1977) Sorptive behavior of trace metals on fly ash in aqueous systems. Environ Sci Technol 11:1096–1100

Tripathi RD, Vajpayee P, Singh N, Rai UN, Kumar A, Ali MB, Kumar B, Yunus M (2004) Efficacy of various amendments for amelioration of fly ash toxicity: growth performance and metal composition of *Cassia siamea* Lamk. Chemosphere 54:1581–1588

Vaidya C (2009) Urban issues, reforms and way forward India. Working paper no. 4/2009-DEA. Department of Economic Affairs, Ministry of Finance, Government of India

Wang HB, Zhang ZX, Li H, He HB, Fang CX, Zhang AJ (2011) Characterization of metaproteomics in crop rhizospheric soil. J Proteome Res 10:932–940

Wheatley AD, Sadhra S (2004) Polycyclic aromatic hydrocarbons in solid residues from waste incineration. Chemosphere 55:743–749

Wong MH, Wong JWC (1986) Effects of fly ash on soil microbial activity. Environ Pollut Ser A 40:127–144

Wong JWC, Wong MH (1987) Co-recycling of fly ash and poultry manure in nutrient-deficient sandy soil. Resour Conserv 13:291–304

Wong JWC, Wong MH (1990) Effects of Fly ash on Yields and Elemental Composition of Two Vegetables, *Brassica parachinensis* and *B. chinensis*. Agric Ecosyst Environ 30:251–264

Yeledhalli NA, Prakash SS, Gurumurthy SB, Ravi MV (2007) Coal fly ash as modifier of physicochemical and biological properties of soil. Karnataka J Agric Sci 20(3):531–534

Zargar S, Nazir M, Cho K, Kim D, Jones O, Sarkar A, Agrawal SB, Shibato J, Kubo A, Jwa N, Agrawal G, Rakwal R (2011) Impact of climatic changes on crop agriculture: OMICS for sustainability & next generation crops. In: Benkeblia N (ed) Sustainable agriculture and new biotechnologies. Taylor & Francis, London, pp 453–478

Oil Palm Biomass as an Adsorbent for Heavy Metals

Mohammadtaghi Vakili, Mohd Rafatullah, Mahamad Hakimi Ibrahim, Ahmad Zuhairi Abdullah, Babak Salamatinia, and Zahra Gholami

Contents

1. Introduction 62
2. Commercial Adsorbents 64
   - 2.1 Activated Carbon 64
   - 2.2 Activated Alumina 64
   - 2.3 Zeolite 65
   - 2.4 Silica Gel 65
3. Agricultural-Waste Adsorbents 65
4. Oil Palm Biomass: Potential Heavy-Metal Adsorbents 69
   - 4.1 Unmodified Oil Palm Biomass 70
   - 4.2 Modified Oil Palm Biomass 72
5. Conclusions 76
6. Summary 80
References 80

M. Vakili • M. Rafatullah (✉) • M.H. Ibrahim
School of Industrial Technology, Universiti Sains Malaysia, 11800 Penang, Malaysia
e-mail: mohd_rafatullah@yahoo.co.in; mrafatullah@usm.my

A.Z. Abdullah • Z. Gholami
School of Chemical Engineering, Universiti Sains Malaysia,
14300 Nibong Tebal, Penang, Malaysia

B. Salamatinia
School of Engineering, Monash University Sunway Campus,
Jalan Lagoon Selatan, 46150 Bandar Sunway, Selangor, Malaysia

1 Introduction

In recent decades, increases in the world's population, unplanned urbanization, industrialization, agricultural activities, and expanded use of chemicals, has contributed to environmental contamination via emission of wastes and pollutants. Wastes (both inorganic and organic) that are produced by human activities have resulted in high volumes of contaminated water, contact with or consumption of which poses health threats to living organisms, including humans (Ahmad et al. 2010, 2012).

Among inorganic pollutants, heavy metals are hazardous pollutants of wastewaters that have become a serious public health concern (Demirbas et al. 2006). Heavy metals harm flora and fauna because they are both toxic and stable; moreover, some of these metals can accumulate in living organisms (Das et al. 2008). The most significant toxic metal ions that pose risks to humans and the environment include Cr, Cu, Pb, Hg, Mn, Cd, Ni, Zn, and Fe (Chatterjee et al. 2010). Duruibe et al. (2007) reported that heavy metals cause adverse health effects, such as gastrointestinal disorders, diarrhea, stomatitis, tremors, hemoglobinuria, ataxia, paralysis, vomiting, and convulsions, although each of these heavy metals exhibits its specific toxicity profile. Wastewater generated from various industrial activities such as battery manufacturing (Ahmaruzzaman 2011), ceramics production (Khraisheh et al. 2004), metal refineries (Chandra Sekhar et al. 2004), pulp and paper production (Sthiannopkao and Sreesai 2009), rubber and plastics manufacture (Srivastava and Majumder 2008), electroplating (Sekomo et al. 2012), smelting (Fu et al. 2012), mining (Ying and Fang 2006), mineral processing and extractive metallurgy (Ahluwalia and Goyal 2007) and metal surface treatment (Karvelas et al. 2003) are contaminated with one or more of these toxic ions. The quantity of these heavy metals that exists in effluents released into the natural environment is often higher than the acceptable level. Hence, heavy metals should be removed or their quantities reduced from effluents by suitable treatment methods before they are discharged into the environment. The industrial sources and health risks of commonly utilized heavy metals are listed in Table 1.

Different treatment methods have been applied to remove heavy metals from wastewaters. Among the common methods are the following: ion exchange (Xing et al. 2007), coagulation/flocculation (Chafi et al. 2011), chemical precipitation (Kurniawan et al. 2006), electrochemical reaction (García-Gabaldón et al. 2006), electro-dialysis (Mohammadi et al. 2004), physisorption (Chen et al. 2012), biosorption (Tsekova et al. 2010), and membrane filtration (Barakat and Schmidt 2010). Each of these methods has been applied to decrease the concentrations of detrimental metal ions in wastewaters. Moreover, each of the methods exhibit limitations, such as high capital or operating costs, low efficiency, and disposal of excess sludge, whereas some of these methods are inappropriate for use by small-scale industries (Kobya et al. 2005).

Ideriah et al. (2012), Ahmed Basha et al. (2008) and Al Aji et al. (2012) studied the advantages and disadvantages of some of these methods, and discovered that precipitation methods are cost effective, but produce high amounts of precipitate

Table 1 Sources of environmental contamination by several heavy metals and their toxic effects

Heavy metals	Sources	Health risks
Lead	Lead batteries, paint, oil, metal, phosphate fertilizer, electronics, wood production, some petrol types, explosive manufacturing, mining activity, automobile emissions, sewage wastewater, sea spray, insecticides, plastic industries, food, beverages, ointments and medicinal concoctions (Khalid et al. 2007)	Dysfunction of kidneys, reproductive system, liver, brain and central nervous system. Reduction in hemoglobin formation, mental retardation, infertility and abnormalities in pregnant women. Anemia, headache, chills, diarrhea, poisoning (Karvelas et al. 2003)
Cadmium	Cadmium–nickel batteries, phosphate fertilizers, pigments, stabilizers, alloys, and electroplating industries (Mortaheb et al. 2009)	Renal disturbances, lung insufficiency, bone lesions, cancer, hypertension (Sankararamakrishnan et al. 2007)
Copper	Mining operations, tanneries, electronics, electroplating, petrochemical industries, and textile mill products (Kazemipour et al. 2008)	Abdominal pain, nausea, vomiting, headache, lethargy, diarrhea, tachycardia, respiratory difficulties, hemolytic anemia, gastrointestinal bleeding, liver and kidney failure and death (Akar et al. 2009)
Mercury	Refineries, coal-fired power plants, mining, chloralkali plants utilizing the Hg-cell process, municipal wastewaters (Urgun-Demirtas et al. 2012)	Neurological and renal disturbances, mental dysfunction, impairment of the nervous system and pulmonary systems and kidney function, and cause chest pain and dyspnea (Zahir et al. 2005)
Manganese	Steel industries, dry battery cells and electrical coils, mining and smelting, pigments and paints, and ceramics (Li et al. 2010)	Damage to brain, liver, kidneys and nervous system (Silva et al. 2010)
Nickel	Stainless steel, super alloys, metal alloys, coins, batteries (Vieira et al. 2010)	Gastrointestinal distress like nausea, vomiting, diarrhoea, damage to lungs and kidney, and cause pulmonary fibrosis, renal edema, and skin dermatitis (Akhtar et al. 2004)
Zinc	Mining, tanneries, painting, car radiator manufacturing, agricultural sources, electroplating, galvanizing plants (Abdelwahab et al. 2013)	Cause abdominal pain, nausea, vomiting, and diarrhea (Pereira et al. 2010)

sludge that requires further treatment. Ion exchange and reverse osmosis efficiently remove heavy metal ions (by approximately 90–95%), but the materials and operational procedures are expensive, and operational problems are often encountered. Electrolysis is an expensive method and requires high energy levels. Commercial activated carbon (CAC) can be applied to remove heavy metals via adsorption, but these adsorbents are very expensive.

More cost-effective and efficient methods and substances are needed to remove heavy metals. Among treatment strategies, adsorption is regarded to be an effective and preferable method for removing heavy metal ions from wastewater, because this method is cost effective and produces high-quality effluent (Oluyemi et al. 2012; Rafatullah et al. 2010; Salleh et al. 2011). Adsorption is a separation process, in which the amount of chemical components being collected (adsorbate) are increased at the surface of a solid (adsorbent) (Yadla et al. 2012). This adsorption process incorporates both physical and chemical actions that involve van der Waals forces, or other actions between an adsorbate and an adsorbent (Wang et al. 2009). Adsorption can function in solid or liquid matrices, and certainly can be used to remove heavy metal ions from polluted aqueous solutions. Adsorption is preferred over other methods because it is rapid, conveniently designed and operated, impenetrable to toxic contaminants, and does not produce hazardous by-products (Qiu et al. 2009). Adsorption is often applied to clean effluents by using low-cost materials.

In this review, we describe the different methods that are used to eliminate heavy metals from wastewaters by using oil palm biomass as a form of low-cost adsorbent.

2 Commercial Adsorbents

The nature and type of adsorbent are important parameters that influence adsorption efficiency. Some of the prominent substances that are commonly used as commercial adsorbents are activated carbon (Mohan and Pittman 2006), activated alumina (Mahmoud et al. 2010), silica gel (Najafi et al. 2011), and zeolite (Egashira et al. 2012). Below, we describe the characteristics of these important adsorbents.

2.1 Activated Carbon

Activated carbon is efficient, adsorbs many chemicals, and is an adsorbent that is particularly important for wastewater treatment (Yin et al. 2008a, b). Activated carbon is produced by dehydration and carbonization in the presence of heat and in the absence of oxygen. Activated carbon contains tiny pores with a large surface area (300–4,000 m^2/g). Although activated carbon is put to many uses, it does possess some limiting features: utilizing it entails high cost, requires regeneration after adsorption, and it loses adsorption capability after regeneration (Igwe and Abia 2006; O'Connell et al. 2008; Rafatullah et al. 2013).

2.2 Activated Alumina

Activated alumina is produced by thermally treating hydrous alumina granules. Hydroxyl groups are forced to leave, producing a porous solid structure of activated alumina that has a large surface area (200–300 m^2/g). Activated alumina possesses a surface area that makes it appropriate for removing heavy metals from aqueous solutions, and absorbing organic liquids (e.g., kerosene, gasoline, and oil) from water (Ku and Chiou 2002; Singh and Pant 2004).

2.3 Zeolite

Zeolites are hydrated porous aluminosilicate minerals. These minerals are naturally created from the changes occurring in glass-rich volcanic rocks (tuff) in the sea or in playa lake waters. Zeolites are appropriate adsorbents for removing heavy metal ions from wastewaters, because such adsorbents exhibit favorable properties that include the following: high ion exchange capability, molecular sieving, catalysis, and sorption properties (Ji et al. 2012; Wang and Peng 2010).

2.4 Silica Gel

Silica gel, invented in the 1920s, is a concentration of $Si(OH)_4$ in siloxane chains. It is produced in three forms: regular-, intermediate-, and low-density gels with surface areas of 750 m^2/g, 300–350 m^2/g, and 100–200 m^2/g, respectively. Such gels are considered to be suitable adsorbents because they remains stable under acidic conditions, exhibit a rapid adsorption capacity, contain a porous structure that has high surface area, and are non-toxic, non-flammable, and not chemically reactive (Fan et al. 2011; Gübbük et al. 2009).

In general, the use of conventional adsorbents increases costs, particularly when high purity adsorbents are used. Therefore, the use of such adsorbents is not commercially economical, and cost is an important when selecting adsorbents. Generally, an adsorbent is regarded to be inexpensive when it is readily available, is environmentally friendly and is cost-effective. Hence, rather than using high-cost adsorbents, researchers are encouraged to produce and use inexpensive adsorbents that are based on natural by-products, such as agricultural wastes, when possible (Bailey et al. 1999; Khan et al. 2008).

In this review, we have searched and summarized the literature that addresses the use of palm oil biomass as a low-cost adsorbent for removing heavy metal contaminants from wastewaters.

3 Agricultural-Waste Adsorbents

Ho (2003) investigated agro-based waste materials as resources to both produce new adsorbents and to modify currently used ones. Previous studies (Basso et al. 2002; Hashem 2007) have demonstrated that agricultural wastes absorb heavy metal ions and can be used as low-cost adsorbents in wastewater treatment. Such wastes have been used for adsorption tasks because they offer several advantages: they are readily available and exist in abundance, they are cost effectiveness, renewable, require less processing time, offer suitable adsorption capability, are selective for heavy metals, and can easily be regenerated (Elizalde-González et al. 2008). Examples of agricultural or related biomass products that can be used in adsorption applications are: peanut skins (Asubiojo and Ajelabi 2009), hazelnut shells (Bulut and Tez 2007a), peanut hulls (Hashem et al. 2005), corn cobs (Sun and Webley 2010), flamboyant pods (Vargas et al. 2010), coconut husks (Tan et al. 2008), Gular fruits (Rao and Rehman 2010), olive stones (Aziz et al. 2009), sawdust (Bulut and Tez 2007b), and chestnut shells (Vázquez et al. 2009).

Saeed et al. (2005) evaluated the efficiency of papaya wood as an adsorbent to remove heavy metals. The percentages of heavy metals removed within 60 min from a solution containing 10 mg/L of Cu (II), Cd (II), and Zn (II) at pH 5 were 97.8%, 94.9%, and 66.8%, respectively. Babarinde et al. (2006) reported the potential of maize leaves for removing Pb ions from wastewater. Agarwal et al. (2006) investigated the efficiency of *Tamarindus indica* seeds, crushed coconut shells, almond shells, groundnut shells, and walnut shells as inexpensive adsorbents for removing Cr (VI). Among these materials, the Cr (VI) sorption capacity of *T. indica* seed was higher than that of the others; crushed coconut shell exhibited the lowest sorption capacity. Abu Al-Rub (2006) studied the effectiveness of palm tree leaves for removing Zn ions from wastewater and found that sorption by Zn was rapid; 90% of Zn was adsorbed in approximately 10 min. Amarasinghe and Williams (2007) investigated the adsorption of Pb and Cu ions from aqueous solutions by using tea waste. They observed that the rate of Pb adsorption was higher than for Cu over a period from 15 to 20 min. Table 2 presents examples of low-cost adsorbents made from various agricultural wastes that are used to remove heavy metals from wastewater.

In general, agricultural wastes are composed of basic components (e.g., cellulose, hemicellulose, and lignin) that contain various functional groups (Amarasinghe and Williams 2007). Lignocellulosic materials are composed of β-D-glucopyranose units, which is one of the most important components of plant cell walls. Each β-D-glucopyranose units contain one primary hydroxyl group and two secondary hydroxyl groups that are commonly involved in chemical reactions. Functional groups present in lignocellulosic materials bind heavy metals by donation of an electron pair from these groups to form complexes with the metal ions in solution (Demirbas 2008). However, the adsorption capacity and physical stability of unmodified lignocellulosic materials are not suited to adsorbing heavy metals. To improve the adsorption capacity for metals, and to enhance metal ion binding, researchers chemically modify these lignocellulosic materials by integrating them with other sources of functional groups in ways that alter their surface characteristics (Mahmoud et al. 2010).

Table 2 Performance parameters of agricultural waste adsorbents that are used for removing heavy metals

	Adsorbent			Adsorbate		Adsorption conditions				Adsorption capacity (mg/g)	References
S. No.	Agricultural waste	Particle size	Dosage	Metals	Concentration (mg/L)	pH	Contact time (min)	Agitation speed (rpm)	Temp. (°C)		
1.	Coconut shell Neem leaves Hyacinth roots Rice straw	250–350 μm	10 g/L	Cu(II)	25	6	300	–	–	19.8886 17.4886 21.7959 18.3519	Singha and Das (2013)
2.	Mushroom biomass	0.1 mm	10 g/L	Cu(II)	30–100	5	30	150	20	0.664	Ertugay and Bayhan (2010)
3.	Potato peel	0.2 mm	1.0 g/100 mL	Cu	150	6	120	150	30	0.3877	Aman et al. (2008)
4.	Waste tea fungal biomass	1–2 mm	0.5 g/L	Cu(II)	254	4	360	200	25	4.64	Razmovski and Šćiban (2008)
5.	Sugar beet pulp	250 μm	0.1 g /100 mL	Cu (II)	250	4	600	150	25	28.5	Aksu and İşoğlu (2005)
6.	Lemon peel	0.10–0.07 mm	10.0 g/L	Cr(VI)	0–1,000	6	600	200	25	22	Bhatnagar et al. (2010)
7.	Parthenium hysterophorus weed	68.5 μm	0.1 g/100 mL	Cr(VI)	10–50	1	420	200	20	24.5	Venugopal and Mohanty (2011)
8.	Pomegranate husk	≤0.063 mm	0.3 g/100 mL	Cr(VI)	75–150	1	180	200	25	35.2	Nemr (2009)
9.	Maize bran	<178 μm	–	Cr(VI)	200	2	180	125	40	312.52	Hasan et al. (2008)
10.	Tea factory waste	0..5–0.25 mm	10 g/L	Cr(VI)	400	2	60	360	60	54.65	Malkoc and Nuhoglu (2007)
11.	Grapefruit peel	355 μm	4 g/L	Ni	10–200	5	60	180	–	46.13	Torab-Mostaedi et al. (2013)
12.	Sugarcane bagasse	<1 mm	1 g/100 mL	Ni (II)	10–200	5	120	300	25	2	Alomá et al. (2012)
13.	Moringa oleifera bark	–	0.2 g/50 mL	Ni	20–200	6	120	300	25	30.38	Reddy et al. (2011)
14.	Acacia leucocephala bark	–	0.1 g/100 mL	Ni	50–200	5	240	150	30	294.1	Subbaiah et al. (2009)
15.	Alternanthera philoxeroides biomass	<125 μm	0.25 g/50 mL	Ni (II)	100	6	300	200	Room temp.	9.73	Wang and Qin (2006)
16.	coconut (Cocos nucifera L.) coir dust	50 μm	5 g/100 mL	Zn	5–200	7.5	60	–	30	17.857	Israel and Eduok (2012)
17.	Sulfured orange peel	0.45 mm	50 mg/10 mL	Zn	50	5–6	120	120	30	80	Liang et al. (2011)
18.	Coffee husks	–	1 g/50 mL	Zn	50–100	4	–	100	25	5.6	Oliveira et al. (2008)
19.	Alternanthera philoxeroides biomass	<125 μm	0.25 g/50 mL	Zn (II)	100	6	300	200	Room temp.	18.57	Wang and Qin (2006)
20.	Rice husk ash Neem bark	0.297–0.400	10 g/L	Zn	25	5	240 180	–	30	14.30 13.29	Bhattacharya et al. (2006)

(continued)

Table 2 (continued)

	Adsorbent			Adsorbate			Adsorption conditions				Adsorption capacity (mg/g)	References
S. No.	Agricultural waste	Particle size	Dosage	Metals	Concentration (mg/L)	pH	Contact time (min)	Agitation speed (rpm)	Temp. (°C)			
21.	Sulfured orange peel	0.45 mm	50 mg/10 mL	Pb	100	–	120	120	30		164	Liang et al. (2011)
22.	Olive tree pruning waste	<1.000 mm	10 g/L	Pb	10	5	120	–	25		26.24	Blázquez et al. (2011)
23.	Moringa oleifera tree leaves	0.6–0.85 mm	0.40 g/100 mL	Pb	50	5	120	200	40		209.54	Reddy et al. (2010)
24.	Orange peel	–	0.050 g/25 mL	Pb	50	5.5	180	120	–		476.1	Feng et al. (2011)
25.	Pine bark	150–355 μm	50 mg/10 mL	Pb	50–1,000	4	240	400	–		76.8	Gundogdu et al. (2009)
26.	Grapefruit peel	355 μm	4 g/L	Cd (II)	10–200	5	60	180	–		42.09	Torab-Mostaedi et al. (2013)
27.	Areca catechu	200 μm	0.5 g/100 mL	Cd (II)	20	6	30	120	29		10.66	Chakravarty et al. (2010)
28.	Banana peel	0.250 mm	30 g/L	Cd (II)	50	3	20	100	25		5.71	Anwar et al. (2010)
29.	Mungbean husk	1.0–2.0 mm	0.5 to 100 m/L	Cd (II)	50	5	60	150	25		35.41	Saeed et al. (2009)
30.	Parthenium	0.104–0.152 mm	0.5 g/50 mL	Cd (II)	10–100	4	60	100	20		27	Ajmal et al. (2006)
31.	Hazelnut hull	–	0.5 g/100 mL	Fe(III)	20–60	3	60	–	30		13.59	Sheibani et al. (2012)
32.	Orange peel	0.841 mm	0.1 g/100 mL	Fe(III)	30	3	360	–	Room temp.		18.19	Lugo-Lugo et al. (2012)
33.	Green tomato husk	0.075–0.150 mm	100 mg/10 mL	Fe(III)	10	6	–	–	20		19.83	García-Mendieta et al. (2012)
34.	Tamarind bark	–	2 g/50 mL	Fe(III)	20–120	2.5	180	200	25		11.75 7.87	Prasad and Abdullah (2009)
	Potato peel											
35.	Bengal gram husk	–	1 g/100 mL	Fe(III)	20–500	2.5	200	100	–		72.16	Ahalya et al. (2006)
36.	Sugarcane bagasse	–	5 g/L	Hg	76	4	180	700	30		35.71	Khoramzadeh et al. (2012)
37.	Gum karaya (Sterculia urens)	180–300 μm	1 g/L	Hg	10	6	60	200	25		62.5	Vinod et al. (2011)
38.	Garlic (Allium sativum L.) powder	0.02 mm	12.5 g/L	Hg	200×10⁻³	–	360	–	Room temp.		0.6497	Eom et al. (2011)
39.	Peat moss	1.0 m	0.125 g/25 mL	Hg	40–523	6	–	–	25		98.94	Bulgariu et al. (2009)
40.	Leaves of castor tree	125–150 μm	0.25 g/100 mL	Hg	5–100	5.5	120	–	Room temp.		37.2	Al Rmalli et al. (2008)
41.	Green tomato husk	0.075–0.150 mm	100 mg/10 mL	Mn	10	6	–	–	20		15.22	García-Mendieta et al. (2012)
42.	Maize stalks	150 μm	0.1 g/25 mL	Mn	40–1,000	5	90	100	35		16.61	El-Sayed et al. (2011)
43.	Pecan nutshell	250 μm	5.0 g/L	Mn	10–1,000	5.5	360	–	25		85.9	Vaghetti et al. (2009)
44.	Black carrot residues	0.250 mm	–	Mn	1,000	5.5	360	–	20		3.871	Güzel et al. (2008)

4 Oil Palm Biomass: Potential Heavy-Metal Adsorbents

Oil palm (*Elaeis guineensis*) biomass is an important and low-cost agricultural waste that exhibits adsorption potential adequate to eliminate heavy metal ions from wastewater (Ibrahim et al. 2010; Ahmad et al. 2011). Oil palm is a tropical tree that originated from Africa. This species has geographically been spread to regions of 42 tropical countries in Africa, the Americas, and Asia. Oil Palm is worldwide covers approximately 27 million acres. Oil palm has been traditionally regarded as an important industrial crop, because it was also utilized for food, in medicine, in woven materials, and in wine over the past 5,000 years. At present, oil extracted from oil palm is used in cooking, cosmetics, pharmaceuticals, and as a bio-fuel (Mohammad et al. 2012). Furthermore, palm oil is one of the largest vegetable oil sources in the world and is a significant economic crop in tropical areas of Africa, America, and Asia, particularly in Southeast Asian countries, such as Indonesia and Malaysia (Kalinci et al. 2011).

Malaysia and Indonesia are among the largest producers of palm oil in the world, and produce approximately 85% of the world's total palm oil (Malaysia 41% and Indonesia 44%). The palm oil industry in Malaysia has expanded rapidly during the past 25 years. This expansion increased the total planted area of oil palm trees from 3.87 million ha in 2004 to 4.17 million ha in 2006 (Sulaiman et al. 2009). In addition, the amount of palm oil produced has increased from 2.5 million tons in 1980 to 17.8 million tons in 2009. Despite growth in area planted, and the oil high production, environmental concerns are increasing about the accumulation of huge quantities of produced biomass wastes (Rupani et al. 2010). Annually, approximately 184 million tons of palm oil residue worldwide, and 53 million tons of oil palm tree residue in Malaysia are generated; these amounts are increasing by ~5% annually (Mohammed et al. 2011).

Large amounts of several components of oil palm biomass are generated and utilized for various purposes. These components include oil and lignocellulosic materials, such as palm pressed fibers (PPF), kernel shells, empty fruit bunch (EFB), oil palm frond (OPF), oil palm trunks, oil palm bark (OPB), palm kernel cake, and palm oil mill effluent (POME) from palm oil production (Uemura et al. 2011). Lignocellulosic oil palm biomass is rich in carbohydrates and contains organic compounds such as cellulose, hemicelluloses and lignin that have numerous natural polymeric materials containing different functional groups that absorb heavy metal ions (Mahmoud et al. 2010). In Table 3, we depict the chemical composition of palm oil biomass.

Table 3 Chemical composition of oil palm biomass

Component	Chemical composition				
	EFB	Frond	Fiber	Trunk	Shell
Cellulose (%)	49.6	25.08	47.6	37.14	27.7
Hemicellulose (%)	18	24.06	25.7	31.8	21.6
Lignin (%)	21.2	18.46	14.1	22.3	44
Ash (%)	2	11.66	1.5	4.3	2.1

Oil palm biomasses can be converted to high-value by-products that can be used as energy sources, erosion control products, soil conditioner, animal feed, fertilizers, as well as in the furniture- and paper-making industries (Radzi bin Abas et al. 2004). Moreover, as we have explained above, palm oil biomass can serve to adsorb heavy metal ions from wastewater.

4.1 Unmodified Oil Palm Biomass

Ho and Ofomaja (2005) studied the kinetics and thermodynamics of Pb ion sorption from aqueous solutions of palm kernel fiber, and discovered that the kinetics followed a pseudo-second-order mechanism. Palm kernel fiber adsorbs Pb ions from aqueous solutions via a spontaneous and endothermic process. The activation energy and equilibrium sorption capacity of Pb ions on palm kernel fiber were determined as 13.5 kJ/mol and 49.9 mg/g at 65 °C, respectively. Salamatinia et al. (2007) assessed the sorption capacity of unmodified OPB, OPF, and EFB for Zn and Cu removal from wastewater. In this study, experiments were conducted in a batch system with 250 mL Cu and Zn solutions at 100 mg/L, using between 0.5 and 1.0 g of adsorbent. OPB, OPF, and EFB adsorbed Cu ions more efficiently than did Zn ions. The sorption capacities of the Zn ions by OPF and EFB were 51.5% and 46.0%, respectively. The Cu sorption capacities of OPF and EFB were 54% and 56.5%, respectively. OPB exhibited the lowest rate of Cu ion removal. Hossain et al. (2012) investigated the removal of Cu from water and wastewater by using untreated palm oil fruit shells as the adsorbent. The raw materials were washed, dried, and ground into powder (<75 mm). Results were that the equilibrium sorption capacity of Cu ranged between 28 and 60 mg/g at room temperature at pH 6.5. Palm oil fruit shells effectively acted as bio adsorbents and eliminated Cu ions from the tested wastewater. Chong et al. (2012) studied the application of oil palm shell as a constructed wetland medium and adsorbent to remove Cu (II) and Pb (II). Results indicate that oil palm shell can be used as filter bed media and can be applied in constructed wetlands to eliminate heavy metals, even without agitation. The sorption capacities determined for this adsorbent were respectively 1.756 and 3.390 mg/g for Cu (II) and Pb (II) ions.

Although unmodified biomass have advantages as adsorbents, they also cause certain problems. Such problems include low adsorption capacity, increased chemical oxygen demand (COD) and biological chemical demand (BOD), and increased total organic carbon (TOC) from release of soluble organics within the biomass. These effects of unmodified biomass adsorbents decrease the oxygen content of water and endanger aquatic life (Peng and Gun 2010). To overcome these disadvantages, and to improve adsorption properties, researchers have sought ways to modify these biomass wastes before using them as adsorbents. Modification is generally designed to improve sorption capacity by creating a charged surface and by increasing the heavy-metal-ion binding capacity (Tijani 2011). In Table 4, we summarize what effects of several unmodified oil palm biomass types have on heavy metal adsorption parameters.

Table 4 Performance parameters of unmodified oil palm biomass-based adsorbents for removing heavy metals

S. no.	Adsorbent Oil palm biomass	Particle size	Dosage	Adsorbate Metals	Concentration (mg/L)	Adsorption conditions pH	Contact time (min)	Agitation speed (rpm)	Temp. (°C)	Adsorption capacity (mg/g)	References
1.	Natural oil palm pressed fibers	250–500 mm	0.1 g/25 mL	Cu	5–25	6	120	250	Room temp.	2.41	Low et al. (1996)
2.	Palm pressed fibers	0.30–0.85 mm	–	Cu, Ni	$50-100 \times 10^{-3}$	6	–	–	–	–	Tan et al. (1996)
3.	Oil palm bark	–	0.5 g/250 mL	Cu	100	–	180	150	25	8.3	Salamatinia et al. (2007)
				Zn						6.3	
				Cu						8.6	
			1 g/250 mL	Zn						6.4	
	Oil palm frond		0.5 g/250 mL	Cu						13.8	
				Zn						13	
			1 g/250 mL	Cu						13.5	
				Zn						12.9	
	Empty fruit bunch		0.5 g/250 mL	Cu						13.0	
				Zn						13.2	
			1 g/250 mL	Cu						14.1	
				Zn						12.3	
4.	Oil palm leaves	250–500 μ	0.5 g/50 mL	Cu	1–100	6	240	125	30	11.22	Sulaiman et al. (2010)
5.	Palm oil fruit shells	<75 μm	0.5 g/100 mL	Cu	10	6.5	600	120	Room temp.	59.502	Hossain et al. (2012)
6.	Bornean oil palm shell	6.5–8 mm	1 g/100 mL	Cu	10	4.1	480	150	–	1.756	Chong et al. (2013)
				Pb						3.390	
7.	Palm kernel fiber	50–60 μm	1 g/400 mL	Pb	120	5	200	65		49.9	Ho and Ofomaja (2005)

4.2 Modified Oil Palm Biomass

4.2.1 Chemical Modification

Results have shown that chemically modifying biomass improves heavy metal removal and sorption capacity. Biomass can be modified by treating it with different chemical agents (e.g., alkalis, acids, organic compounds, etc.). Such chemical modification increases the level of metal uptake by releasing certain soluble organic compounds within the biomass (Abdullah et al. 2009).

Tan et al. (1993) removed Cr (VI) from wastewater in batch and column systems by treating PPF and coconut husk (CHF). The substrates, after boiling in distilled water, were treated stepwise with 1.5 M NaOH, distilled water, 2 M HNO_3 and distilled water. In the batch system, Cr (VI) was efficiently removed at pH ranges of 1.5 to 3 and 1.5 to 5 by PPF and CHF, respectively. The sorption capacities of PPF and CHF are 14 and 29 Cr/g substrate at pH 2.0, respectively. In the column system, PPF and CHF removed Cr (VI) ions from wastewater at various flow rates and bed depths. These substrates were also used as barriers in landfills to prevent Cr (VI) from leaching. Low et al. (1996) showed that the amount of Cu removed from wastewater by dye-treated oil PPF was higher than that by an untreated PPF. The results obtained from batch and column tests indicated that the use of PPF to remove Cu (II) ions was efficient. The sorption capacities of natural and dye-coated PPFs were 2.41 and 7.71 mg/g, respectively; the sorption capacity of these adsorbents was dependent on pH and Cu ion concentration in the solution. Further, Abia and Asuquo (2008) compared the sorption capacities of modified and unmodified oil palm fruit fibers as adsorbents to remove Pb and Cd ions from wastewater. Chemically modified adsorbents (treated with 0.3 HNO_3) increased the sorption capacities of Pb and Cd to 5.579 and 7.980 mg/g, respectively.

Salamatinia et al. (2006) modified OPF by applying a chemical pre-treatment and then using it to remove Zn and Cu ions from wastewater. Different pre-treatments (e.g., acid, base, steam, and reactive dye) were used to improve the sorption capacity of OPF. OPF treated with a base (1.0 M NaOH) for 45 min at 25 °C showed the highest improvement in heavy metal removal capacity (64%). The effect of base concentration was greater than the effect of treatment time. Abia and Asuquo (2007) compared the effects of unmodified and mercaptoacetic acid-modified oil palm fruit fiber to sorb Cd (II) and Cr (III) from wastewater. The sorption equilibrium of both metals was reached after 1 h. The modified adsorbent exhibited better removal efficiency, because the thiolation reaction influenced adsorbent behavior. In addition, the rate of Cr (III) ion removal by both adsorbents was higher than that of Cd (II) ion removal. The intraparticle diffusion rate constants of Cd (II) ion were 62.04, 67.01, and 71.43 min^{-1}; for Cr (III) these values were 63.41, 65.79, and 66.25 min^{-1}. Akaninwor et al. (2007) analyzed the efficacy of thioglycolic-modified oil palm fiber to remove Fe, Zn, and Mg ions from wastewater. In Southern Point tests, the highest sorption capacities for Fe (II), Zn (II), and Mg (II) were respectively 83.6%, 75.6%, and 50.8%; in Northern Point tests, the highest sorption capacities for Fe

(II), Zn (II), and Mg (II) were 79.1%, 78.3%, and 77.5%, respectively at pH 6. Therefore, the removal efficiency of these ions was influenced by pH and ionic size. The volume of adsorbed Fe (II) was the highest, followed by Zn (II) and Mg (II).

Abdullah et al. (2009) improved heavy metal sorption by treating OPF with 0.1 and 1.1 M NaOH for a maximum of 5 h. The maximum sorption capacities of Zn and Cu removal were 61.5% and 64.0%, respectively, under the following optimum conditions: 1.0 g of OPF treated with 1.0 M NaOH in 250 mL of 100 mg/L Zn and Cu solutions for 45 min. NaOH treatment improved the sorption capacity by increasing the rate of metal binding. Haron et al. (2009) used hydroxamic acid-modified EFB for Cu (II) sorption. The raw material was grafted by treatment with polymethylacrylate and then was treated with hydroxylammonium chloride, thereby decreasing the intensity of the adsorption band from 1,734 cm^{-1} to 1,640 cm^{-1}. An absorption band was also obtained at 1,568 cm^{-1}, which corresponds to the N–H amide group. Therefore, a new maximum sorption capacity of 74.1 mg/g was obtained at 25 °C and at pH 4 to 6 by a spontaneous and exothermic process. As a result, hydroxamic acid grafted oil palm empty fruit bunch (PHA-OPEFB) can be used as an adsorbent to remove Cu (II) from wastewater. In Table 5, we summarize how different heavy metal ions are adsorbed by chemically modified forms of oil palm biomass.

4.2.2 Thermal Modification (Activated Carbon)

Activated carbon is widely used as an adsorbent to eliminate heavy metals from wastewater, because this substance exhibits good adsorption properties as a result of having numerous tiny pores and a large surface area. When choosing adsorbents cost is important, and using activated carbons commercially generally increases adsorption costs. Therefore, utilizing other more cost-effective adsorbents that are environmentally friendly, such as agricultural wastes, have been investigated. As previously mentioned, researchers have investigated oil palm biomasses an alternative adsorbent, because these materials are great sources of high-quality and low-cost activated carbon.

Wan Nik et al. (2006) utilized shell waste from palm oil trees to produce activated carbon as a heavy metal adsorbent. The activated carbon produced by phosphoric acid-treated raw material was used to adsorb Cu, Pb, Cr, and Cd. This treatment decreased the concentration of inorganic elements and increased the surface area of the activated carbon. The optimum Brunet Elmer Teller (BET) surface area (1,058 m^2/g) and pore diameter (20.64 nm) were obtained under the following controlled conditions: 30% phosphoric acid concentration and an activation temperature of 500 °C, with a holding time of 2 h. The adsorption capacities of Cr, Pb, Cd, and Cu were 100%, 99.8%, 99.5%, and 25%, respectively. Issabayeva et al. (2006) analyzed the sorption capacity of Pb from wastewater by using a commercially available palm shell activated carbon. This form of activated carbon can be efficiently used as an adsorbent to remove heavy metals, particularly Pb ions, from wastewater with a high adsorption capacity of 95.2 mg/g at pH 5. The effect of adding malonic acid and boric acid on the sorption capacity of Pb ions was also examined. Boric acid

Table 5 Performance parameters of chemically modified oil palm biomass-based adsorbents for removing heavy metals

S. no.	Adsorbent Oil palm biomass	Particle size	Dosage	Adsorbate Metals	Concentration (mg/L)	pH	Contact time (min)	Agitation speed (rpm)	Temp. (°C)	Adsorption capacity (mg/g)	References
1.	Dye-treated oil palm pressed fibers	250–500 mm	0.1 g/25 mL	Cu	5–25	6	120	250	Room temp.	7.71	Low et al. (1996)
2.	Palm kernel fiber	50–60 μm	1 g/100 mL	Cu	50–250	5.01	60	200	26	–	Ho and Ofomaja (2006a)
3.	Treated oil palm frond	–	1 g/250 mL	Cu, Zn	100	–	30	150	25	–	Abdullah et al. (2009)
4.	Hydroxamic acid modified oil palm empty fruit bunch	100–200 μm	0.1 g/20 mL	Cu	100	4	120	–	25	74.1	Haron et al. (2009)
5.	Palm kernel fiber	50–60 μm	1 g /100 mL	Cu	90.24	5.1	1.5	200	24	20.12	Ofomaja (2010)
6.	Palm pressed fibers	0.30–0.85 mm	0.40 g/L	Cr (VI)	20	2	120	–	–	14	Tan et al. (1993)
7.	Treated oil palm fuel ash	0.5–1.0×10⁻³ mm	–	Cr	–	6	–	300	25	16.11	Chu and Hashim (2003)
8.	Acetic acid modified oil palm fruit fiber	106 μm	0.5 g/L	Cr (III), Cd	50	–	120	–	28	–	Abia and Asuquo (2007)
9.	Oil palm fruit fiber	106 μm	0.5 g/100 mL	Ni, Pb	50	6.2	120	–	28	–	Abia and Asuquo (2006)
10.	Modified oil-palm fibre	106 μm	1 g/50 mL	Zn, Mn, Fe(III)	–	6	60	–	–	–	Akaninwor et al. (2007)
11.	Modified oil palm fruit fiber	106 μm	0.5 g/100 mL	Pb Cd	50	6.2	60	–	28	5.579 7.980	Abia and Asuquo (2008)
12.	Palm kernel fiber	50–60 μm	0.6 g/400 mL	Pb	120	5	20	200	36	47.6	Ho and Ofomaja (2006b)

enhanced the total amount of Pb removed, particularly at pH 5. By contrast, malonic acid decreased adsorption because an aqueous Pb-malonate complex was formed. Iyagba and Opete (2009) used palm kernel shell- and husk-activated carbon as adsorbents in a batch test to remove Cr and Pb from wastewater. The removal rate of Cr and Pb depends on pH, contact time, and adsorbent concentration; the highest removal rates were obtained at an optimum pH of 3 and 5 for Cr and Pb, respectively. Equilibrium times were 90 and 120 min for the activated palm kernel shell and activated palm kernel husk, respectively. The maximum sorption rates for Cr and Pb were 90% and 88%, respectively, and these rates were achieved at an adsorbent loading of 4 g.

Considering adsorbent and method costs as well as adsorption efficiency of heavy metals in industrial wastewater, Nomanbhay and Palanisamy (2005) utilized chitosan-coated acid-treated oil palm shell charcoal to remove Cr ions from polluted industrial wastewater. The adsorption capacity (154 mg Cr/g at 25 °C) of this adsorbent was estimated by using a Langmuir isotherm model under equilibrium conditions. After adsorption was completed, the adsorbent was regenerated with 0.1 M of sodium hydroxide. This adsorbent was technically feasible, environmentally friendly, and highly efficient. Sugawara et al. (2007) used a carbonaceous adsorbent from palm shell to remove Pb^{2+} and Zn^{2+} from wastewater. This adsorbent was prepared by pyrolysis and sulfur impregnation. The pyrolyzed samples with KOH were sulfurized with impregnated H_2S to produce a sulfur-impregnated char exhibiting heavy metal sorption capability. Sulfur impregnation increased sulfur content and enhanced adsorption capacity. Alam et al. (2008) used activated carbon made from empty fruit bunches of oil palm to remove Zn ion from polluted wastewater. The samples were thermally activated at 500, 750, and 1,000 °C for 15, 30, and 45 min. The activated carbon obtained at 1,000 °C for 30 min showed the maximum sorption capacity of 1.63 mg/g, at which 98% of Zn concentration was removed from the wastewater. Wahi et al. (2009) assessed the ability of activated carbon from palm oil EFB to remove Hg, Pb, and Cu from wastewater. The adsorption efficiencies of activated carbon made from EFB for Pb (II), Hg (II), and Cu (II) were 100%, 100% and 25%, respectively. The sorption of these ions by activated carbon of EFB was dependent on the amount of adsorbent and the initial concentration of the metals. Therefore, EFB in the form of activated carbon can be used as an effective adsorbent to remove heavy metals and solve environmental problems caused by high amounts of agricultural wastes.

Granular activated carbon made from palm kernel shell can also be used as an adsorbent to remove Cu, Ni, and Pb ions from industrial wastewater (Onundi et al. 2010). The sorption capacities for Pb, Cu, and Ni were 1.337, 1.581, and 0.130 mg/g, respectively. These values were obtained under the following optimum conditions: pH 5 and 1 g/L of adsorbent. The following equilibrium time was obtained: for Pb, 30 min; for Cu and Ni, 75 min. The proportions of metal ion removal achieved at equilibrium were 100%, 97%, and 55% for Pb, Cu, and Ni: Pb(II) > Cu(II) > Ni(II). Kabbashi et al. (2011) analyzed the adsorption efficiency of an empty-fruit-bunch activated carbon to remove Hg (II) from wastewater. Hg binding was influenced by pH, mixing speed, sorbent concentration and contact time. The sorption capacity of

99.53% was obtained under the following conditions: pH 6.5; mixing speed, 100 rpm; contact time, 70 min; and sorbent concentration, 20 mg. Isa et al. (2008) conducted batch tests with sulfuric acid and heat-treated oil palm fiber to remove Cr(VI) from wastewater. The results showed that the removal efficiency for Cr (VI) was dependent on pH, contact time, initial Cr concentration, and amount of adsorbent used. Oil palm fiber can be used as an inexpensive adsorbent to remove Cr (VI) from wastewater.

Chemical modifications produce increased sorption capacity. Nwabanne et al. (2011) and Nwabanne and Igbokwe (2012) used oil palm empty-fruit-bunch activated carbon and oil-palm-fiber activated carbon in a packed bed column to remove Pb (II) from wastewater. Adsorption efficiency was dependent on initial ion concentration, bed height, and flow rate. Sorption capacity was improved as initial ion concentration and bed height increased, because metals can access more sorption sites under these conditions. By contrast, sorption capacity decreased as flow rate increased, because of decrease time for saturation. Gulnaziya et al. (2012) used commercial untreated palm shell activated carbon (PSAC) and modified PSAC by *Aspergillus niger* and *Bacillus subtilis* to remove Pb ion from wastewater. The experiments were conducted in a batch system at pH 3 to 6 with 20 mg/L to 300 mg/L of Pb. At pH 6, the highest values of Pb uptake were recorded for PSAC-*B. subtilis*, PSAC-*A. niger*, and the original PSAC uptake values were 74, 72, and 65 mg Pb/g, respectively. At pH 3, the lowest uptake values were obtained: 34, 37, and 40 mg Pb/g, respectively. Therefore, biomodification of a PSAC matrix can enhance sorption capacity of Pb ions (90%).

Rahman et al. (2012) assessed the adsorption capacity of chemically-modified activated carbon of palm shell to eliminate Cr, Pb, Cd, and Cu ions from polluted aqueous solutions by using a water filtration column. Palm shells were converted to activated carbon that had a large pore surface area (1,058 m^2/g^{-1}) and a large pore size (20.64 nm diameter) under the following optimum conditions: treatment with 20% H_2SO_4 in solution at 24 h in 30% H_3PO_4 solution, and maintained at 500 °C for 2 h. The adsorption capacities of this adsorbent were 100%, 99.8%, 99.5%, and 25% for Cr, Pb, Cd, and Cu, respectively. In Table 6 we summarize how different heavy metal ions are adsorbed by oil palm biomass carbonaceous adsorbents.

5 Conclusions

The significant increase in production and use of heavy metals in industry has contributed to environmental pollution as a result of the release of high amounts of contaminated water. This increasing heavy metal pollution of waters threatens human health and the environment. Different methods have been used to remove heavy metals from wastewater for the purpose of improving the quality of water that is ultimately discharged to the environment. Although no single method is completely successful in eliminating heavy metals from water, some adsorption solutions produce high quality effluents at relatively low cost. The nature and type of

Table 6 Performance parameters of thermally modified oil palm biomass-based adsorbents for removing heavy metals

	Adsorbent			Adsorbate		Adsorption conditions				Adsorption capacity (mg/g)	References
S. no.	Oil palm biomass	Particle size	Dosage	Metals	Concentration (mg/L)	pH	Contact time (min)	Agitation speed (rpm)	Temp. (°C)		
1.	Oil palm ash	–	2.5 g/L	Ni (II)	40	5	120	200	25	9.9	Chu and Hashim (2003)
2.	Palm shell activated carbon	0.8–1.0 mm	–	Pb	10–700	5	–	150	27	95.2	Issabayeva et al. (2006)
						3				82.0	
3.	Palm oil empty fruit bunch activated carbon	0.5–1.0 mm	1 g/100 mL	Cu Hg Pb	10–20	4.5	–	150	29–31	0.84 52.67 48.96	Wahi et al. (2009)
4.	Activated carbon from palm kernel shell	1.68–2.38 mm	1 g/L	Cu Ni Pb	2.0	5	120	100	27	1.581 0.130 1.337	Onundi et al. (2010)
5.	Modified activated carbon from waste palm shell	2.0 mm	–	Cu Cr Pb Cd	100	–	–	–	–	75.404×10^{-3} – 0.204×10^{-3} 0.455×10^{-3}	Rahman et al. (2012)
6.	Acid-treated oil palm shell charcoal coated with chitosan	100–150 µm	40 g/L	Cr	20	4	180	200	25	154	Saifuddin and Kumaran (2005)
7.	Empty fruit bunch activated carbon	250 µm	10–30 mg/ 50 mL	Hg	0.1	6.5	70	100	–	–	Kabbashi et al. (2011)
8.	Sulphuric acid and heat-treated oil palm fiber	–	0.5 g/100 mL	Cr (VI)	20	1.5	360	350	28	–	Isa et al. (2008)
9.	Palm kernel shell Palm kernel husk	–	2 g/50 mL	Cr, Pb	–	3	90 120	–	–	–	Iyagba and Opete (2009)

(continued)

Table 6 (continued)

S. no.	Adsorbent		Adsorbate		Adsorption conditions					References
	Oil palm biomass	Particle size	Dosage	Metals	Concentration (mg/L)	pH	Contact time (min)	Agitation speed (rpm)	Temp. (°C)	Adsorption capacity (mg/g)	
10.	Palm shell activated carbon	710–850 μm	250 mg/50 mL	Ni	20–750	3	–	180	25	4.5	Yin et al. (2008b)
	Polyethyleneimine-impregnated palm shell activated carbon			Cd		5				6.1	
	Polyethyleneimine-impregnated palm Shell activated carbon			Ni						6	
				Cd						11.9	
	Palm shell activated carbon			Ni						6.1	
				Cd						8.5	
				Ni						9.6	
				Cd						14.1	
11.	Oil palm waste ash	–	3 g/L	Zn	20	6	40	300	25	10.66	Chu and Hashim (2002)
12.	Carbonaceous adsorbent from palm shell	212–355 μm	0.5 g/100 mL	Zn, Pb	157	–	–	–	25	–	Sugawara et al. (2007)
13.	Oil palm empty fruit bunches activated carbon	≤0.5 mm	6 g/L	Zn	10	5.5	120	150	30	1.63	Alam et al. (2008)
14.	Palm shell activated carbon	710–850 μm	250 mg/50 mL	Pb	20–750	3	–	180	25	39	Yin et al. (2008a)
	Polyethyleneimine-impregnated palm shell activated carbon									32.5	
	Palm shell activated carbon					5				66.5	
	Polyethyleneimine-impregnated palm shell activated carbon									53.5	
15.	Palm shell activated carbon	–	–	Pb 1.0 (L/h)	–	5	–	–	–	90.2	Issabayeva et al. (2008)
				Pb 2.0 (L/h)						92.6	
				Pb 0.5 (L/h)		3				22.6	
				Pb 1.0 (L/h)						17.9	
16.	Palm shell activated carbon (PSAC)	1.6–2.0 mm	250 mg/100 mL	Pb	20–300	6	360	150	27	65	Gulnaziya et al. (2012)
	Modified PSAC by *Aspergillus niger*									72	
	Modified PSAC by *Bacillus subtilis*									74	

adsorbent used is critical in influencing the ultimate adsorption efficiency achieved. In general, an adsorbent is considered to be good when it is cost effective, available, environmentally friendly, and does not require a lot of processing. The use of palm oil biomasses as adsorbents to remove heavy metals from contaminated water has been studied by numerous researchers. These adsorbents have specific characteristics that offer several advantages that include: low cost, high absorption capability, environmentally friendly, and biodegradable. If processed appropriately, palm oil biomasses are efficient adsorbents that have extraordinary absorption capability for eliminating heavy metals from waste streams.

In this paper, we have reviewed and compared the adsorption efficiency of several different palm oil biomasses for heavy metals. Increasingly, bio adsorbents like palm oil biomasses are being considered as alternatives to replace conventional adsorbents for removing heavy metals from waste streams. In addition, scientists are working to chemically or structurally modify palm oil biomasses to improve their performance characteristics. Results indicate that such modification can improve sorption capacity by creating a charged surface and by increasing the heavy metal ion binding capacity. Although palm oil biomasses (modified and unmodified) represent good alternatives for replacing commercial adsorbents, additional information on their performance is needed if they are going to be useful for applications at the industrial scale. Developing a multipurpose adsorbent that can remove multiple pollutants from industrial effluents is a reasonable future goal, if the proper research work is undertaken and is successful. From our review, we have concluded that more information is specifically needed in the following areas:

- More complex adsorbents capable of treating industrial wastewater must be investigated.
- Detailed regeneration studies must be performed to enhance the understanding of the economic feasibility of using bio adsorbents such as palm oil biomass. To date, few regeneration studies have been reported. Regeneration studies will determine the reusability of adsorbents made from palm oil biomasses and will contribute to their effectiveness.
- In work performed to date, cost information on oil palm biomasses as adsorbents is seldom addressed or reported in publications. Such cost information is urgently needed. Although modified biomasses can enhance the adsorption of heavy metal ions, the expense of chemicals used and methods of modification also have to be taken into consideration if low-cost adsorbents are to be developed.
- The potential of oil palm biomasses as adsorbents for multi-component pollutants must be assessed. Moreover, these materials must be tested under real industrial effluent conditions. Having such data would significantly assist in moving toward the potential commercial use of biomasses to treat and clean industrial pollution.
- Most researchers have studied oil palm biomass adsorption only in small scale batch processes. Research must now be extended to the pilot-plant scale to better assess oil palm biomass as adsorbents feasible for use at the commercial and industrial scale.

6 Summary

Many industries discharge untreated wastewater into the environment. Heavy metals from many industrial processes end up as hazardous pollutants of wastewaters. Heavy metal pollution has increased in recent decades and there is a growing concern for the public health risk they may pose. To remove heavy metal ions from polluted waste streams, adsorption processes are among the most common and effective treatment methods. The adsorbents that are used to remove heavy metal ions from aqueous media have both advantages and disadvantages. Cost and effectiveness are two of the most prominent criteria for choosing adsorbents. Because cost is so important, great effort has been extended to study and find effective lower cost adsorbents. One class of adsorbents that is gaining considerable attention is agricultural wastes. Among many alternatives, palm oil biomasses have shown promise as effective adsorbents for removing heavy metals from wastewater. The palm oil industry has rapidly expanded in recent years, and a large amount of palm oil biomass is available. This biomass is a low-cost agricultural waste that exhibits, either in its raw form or after being processed, the potential for eliminating heavy metal ions from wastewater. In this article, we provide background information on oil palm biomass and describe studies that indicate its potential as an alternative adsorbent for removing heavy metal ions from wastewater. From having reviewed the cogent literature on this topic we are encouraged that low-cost oil-palm-related adsorbents have already demonstrated outstanding removal capabilities for various pollutants.

Because cost is so important to those who choose to clean waste streams by using adsorbents, the use of cheap sources of unconventional adsorbents is increasingly being investigated. An adsorbent is considered to be inexpensive when it is readily available, is environmentally friendly, is cost-effective and be effectively used in economical processes. The advantages that oil palm biomass has includes the following: available and exists in abundance, appears to be effective technically, and can be integrated into existing processes. Despite these advantages, oil palm biomasses have disadvantages such as low adsorption capacity, increased COD, BOD and TOC. These disadvantages can be overcome by modifying the biomass either chemically or thermally. Such modification creates a charged surface and increases the heavy metal ion binding capacity of the adsorbent.

Acknowledgements The authors acknowledge the research grant provided by the Universiti Sains Malaysia under the Short Term Grant Scheme (Project No. 304/PTEKIND/6312008).

References

Abdelwahab O, Amin NK, El-Ashtoukhy ESZ (2013) Removal of zinc ions from aqueous solution using a cation exchange resin. Chem Eng Res Des 91(1):165–173

Abdullah A, Salamatinia B, Kamaruddin A (2009) Application of response surface methodology for the optimization of NaOH treatment on oil palm frond towards improvement in the sorption of heavy metals. Desalination 244(1):227–238

Abia A, Asuquo E (2006) Lead (II) and nickel (II) adsorption kinetics from aqueous metal solutions using chemically modified and unmodified agricultural adsorbents. Afr J Biotechnol 5(16):1475–1482

Abia A, Asuquo E (2007) Kinetics of Cd^{2+} and Cr^{3+} sorption from aqueous solutions using mercaptoacetic acid modified and unmodified oil palm fruit fibre(elaeis guineensis) adsorbents. Tsinghua Sci Technol 12(4):485–492

Abia A, Asuquo E (2008) Sorption of Pb (II) and Cd (II) ions onto chemically unmodified and modified oil palm fruit fibre adsorbent: Analysis of pseudo second order kinetic models. Indian J Chem Technol 15(4):341–348

Abu Al-Rub FA (2006) Biosorption of zinc on palm tree leaves: equilibrium, kinetics, and thermodynamics studies. Sep Sci Technol 41(15):3499–3515

Agarwal G, Bhuptawat HK, Chaudhari S (2006) Biosorption of aqueous chromium(VI) by Tamarindus indica seeds. Bioresour Technol 97(7):949–956

Ahalya N, Kanamadi ND, Ramachandra TV (2006) Biosorption of Iron (III) from aqueous solutions using the husk of cicer arietinum. Indian J Chem Technol 13:122–127

Ahluwalia SS, Goyal D (2007) Microbial and plant derived biomass for removal of heavy metals from wastewater. Bioresour Technol 98(12):2243–2257

Ahmad T, Rafatullah M, Ghazali A, Sulaiman O, Hashim R, Ahmad A (2010) Removal of pesticides from water and wastewater by different adsorbents: A review. J Environ Sci Health C 28(4):231–271

Ahmad T, Rafatullah M, Ghazali A, Sulaiman O, Hashim R (2011) Oil palm biomass–based adsorbents for the removal of water pollutants-A review. J Environ Sci Health C 29(3):177–222

Ahmad T, Danish M, Rafatullah M, Ghazali A, Sulaiman O, Hashim R, Ibrahim MNM (2012) The use of date palm as a potential adsorbent for wastewater treatment: a review. Environ Sci Pollut Res Int 19(5):1464–1484

Ahmaruzzaman M (2011) Industrial wastes as low-cost potential adsorbents for the treatment of wastewater laden with heavy metals. Adv Colloid Interface Sci 166(1):36–59

Ahmed Basha C, Bhadrinarayana N, Anantharaman N, Meera Sheriffa Begum K (2008) Heavy metal removal from copper smelting effluent using electrochemical cylindrical flow reactor. J Hazard Mater 152(1):71–78

Ajmal M, Rao RAK, Ahmad R, Khan MA (2006) Adsorption studies on parthenium hysterophorous weed: removal and recovery of Cd(II) from wastewater. J Hazard Mater 135(1):242–248

Akaninwor J, Wegwu M, Iba I (2007) Removal of iron, zinc and magnesium from polluted water samples using thioglycolic modified oil-palm fibre. Afr J Biochem Res 1(2):011–013

Akar ST, Akar T, Kaynak Z, Anilan B, Cabuk A, Tabak O, Demir TA, Gedikbey T (2009) Removal of copper(II) ions from synthetic solution and real wastewater by the combined action of dried Trametes versicolor cells and montmorillonite. Hydrometallurgy 97(1–2):98–104

Akhtar N, Iqbal J, Iqbal M (2004) Removal and recovery of nickel(II) from aqueous solution by loofa sponge-immobilized biomass of Chlorella sorokiniana: characterization studies. J Hazard Mater 108(1–2):85–94

Aksu Z, İşoğlu İA (2005) Removal of copper (II) ions from aqueous solution by biosorption onto agricultural waste sugar beet pulp. Process Biochem 40(9):3031–3044

Al Aji B, Yavuz Y, Koparal AS (2012) Electrocoagulation of heavy metals containing model wastewater using monopolar iron electrodes. Sep Purif Technol 86:248–254

Al Rmalli SW, Dahmani AA, Abuein MM, Gleza AA (2008) Biosorption of mercury from aqueous solutions by powdered leaves of castor tree (Ricinus communis L.). J Hazard Mater 152(3):955–959

Alam MZ, Muyibi SA, Kamaldin N (2008) Production of Activated carbon from oil palm empty fruit bunches for removal of zinc. In: Twelfth international water technology conference (IWTC12), Egypt, Alexandria, pp 1–11

Alomá I, Martín-Lara M, Rodríguez I, Blázquez G, Calero M (2012) Removal of nickel (II) ions from aqueous solutions by biosorption on sugarcane bagasse. J Taiwan Inst Chem Eng 43(2):275–281

Aman T, Kazi AA, Sabri MU, Bano Q (2008) Potato peels as solid waste for the removal of heavy metal copper (II) from waste water/industrial effluent. Colloid Surf B 63(1):116–121

Amarasinghe B, Williams R (2007) Tea waste as a low cost adsorbent for the removal of Cu and Pb from wastewater. Chem Eng J 132(1):299–309

Anwar J, Shafique U, Salman M, Dar A, Anwar S (2010) Removal of Pb (II) and Cd (II) from water by adsorption on peels of banana. Bioresour Technol 101(6):1752–1755

Asubiojo O, Ajelabi O (2009) Removal of heavy metals from industrial wastewaters using natural adsorbents. Toxicol Environ Chem 91(5):883–890

Aziz A, Ouali MS, Elandaloussi EH, De Menorval LC, Lindheimer M (2009) Chemically modified olive stone: a low-cost sorbent for heavy metals and basic dyes removal from aqueous solutions. J Hazard Mater 163(1):441–447

Babarinde NA, Babalola JO, Sanni RA (2006) Biosorption of lead ions from aqueous solution by maize leaf. Int J Phys Sci 1(1):23–26

Bailey SE, Olin TJ, Bricka RM, Adrian DD (1999) A review of potentially low-cost sorbents for heavy metals. Water Res 33(11):2469–2479

Barakat M, Schmidt E (2010) Polymer-enhanced ultrafiltration process for heavy metals removal from industrial wastewater. Desalination 256(1):90–93

Basso M, Cerrella E, Cukierman A (2002) Lignocellulosic materials as potential biosorbents of trace toxic metals from wastewater. Ind Eng Chem Res 41(15):3580–3585

Bhatnagar A, Minocha A, Sillanpää M (2010) Adsorptive removal of cobalt from aqueous solution by utilizing lemon peel as biosorbent. Biochem Eng J 48(2):181–186

Bhattacharya A, Mandal S, Das S et al (2006) Adsorption of Zn (II) from aqueous solution by using different adsorbents. Chem Eng J 123(1):43–51

Blázquez G, Martín-Lara M, Tenorio G, Calero M (2011) Batch biosorption of lead (II) from aqueous solutions by olive tree pruning waste: equilibrium, kinetics and thermodynamic study. Chem Eng J 168(1):170–177

Bulgariu L, Ratoi M, Bulgariu D, Macoveanu M (2009) Adsorption potential of mercury (II) from aqueous solutions onto Romanian peat moss. J Environ Sci Health A 44(7):700–706

Bulut Y, Tez Z (2007a) Adsorption studies on ground shells of hazelnut and almond. J Hazard Mater 149(1):35–41

Bulut Y, Tez Z (2007b) Removal of heavy metals from aqueous solution by sawdust adsorption. J Environ Sci 19(2):160–166

Chafi M, Gourich B, Essadki AH, Vial C, Fabregat A (2011) Comparison of electrocoagulation using iron and aluminium electrodes with chemical coagulation for the removal of a highly soluble acid dye. Desalination 281:285–292

Chakravarty P, Sarma NS, Sarma H (2010) Biosorption of cadmium(II) from aqueous solution using heartwood powder of Areca catechu. Chem Eng J 162(3):949–955

Chandra Sekhar K, Kamala C, Chary N, Sastry A, Nageswara Rao T, Vairamani M (2004) Removal of lead from aqueous solutions using an immobilized biomaterial derived from a plant biomass. J Hazard Mater 108(1):111–117

Chatterjee S, Bhattacharjee I, Chandra G (2010) Biosorption of heavy metals from industrial waste water by Geobacillus thermodenitrificans. J Hazard Mater 175(1):117–125

Chen D, Li Y, Zhang J, Li W, Zhou J, Shao L, Qian G (2012) Efficient removal of dyes by a novel magnetic Fe_3O_4/ZnCr-layered double hydroxide adsorbent from heavy metal wastewater. J Hazard Mater 243:152–160

Chong H, Chia P, Ahmad M (2012) The adsorption of heavy metal by Bornean oil palm shell and its potential application as constructed wetland media. Bioresour Technol 130:181–186

Chu KH, Hashim MA (2002) Adsorption and desorption characteristics of zinc on ash particles derived from oil palm waste. J Chem Technol Biotechnol 77(6):685–693

Chu K, Hashim M (2003) Kinetic studies of copper (II) and nickel (II) adsorption by oil palm ash. J Ind Eng Chem 9(2):163–167

Das N, Vimala R, Karthika P (2008) Biosorption of heavy metals—an overview. Indian J Biotechnol 7:159–169

Demirbas A (2008) Heavy metal adsorption onto agro-based waste materials: a review. J Hazard Mater 157(2):220–229

Demirbas A, Sari A, Isildak O (2006) Adsorption thermodynamics of stearic acid onto bentonite. J Hazard Mater 135(1):226–231

Duruibe J, Ogwuegbu M, Egwurugwu J (2007) Heavy metal pollution and human biotoxic effects. Int J Phys Sci 2(5):112–118

Egashira R, Tanabe S, Habaki H (2012) Adsorption of heavy metals in mine wastewater by Mongolian natural zeolite. Procedia Eng 42:54–64

Elizalde-González MP, Mattusch J, Wennrich R (2008) Chemically modified maize cobs waste with enhanced adsorption properties upon methyl orange and arsenic. Bioresour Technol 99(11):5134–5139

El-Sayed GO, Dessouki HA, Ibrahiem SS (2011) Removal of zn(ii), cd(ii) and mn(ii) from aqueous solutions by adsorption on maize stalks. Malayas J Anal Sci 15(1):8–21

Eom Y, Won JH, Ryu J-Y, Lee TG (2011) Biosorption of mercury (II) ions from aqueous solution by garlic (Allium sativum L.) powder. Korean J Chem Eng 28(6):1439–1443

Ertugay N, Bayhan Y (2010) The removal of copper (II) ion by using mushroom biomass (Agaricus bisporus) and kinetic modelling. Desalination 255(1):137–142

Fan H-T, Sun T, Xu H-B, Yang Y-J, Tang Q, Sun Y (2011) Removal of arsenic (V) from aqueous solutions using 3-[2-(2-aminoethylamino) ethylamino] propyl-trimethoxysilane functionalized silica gel adsorbent. Desalination 278(1):238–243

Feng N, Guo X, Liang S, Zhu Y, Liu J (2011) Biosorption of heavy metals from aqueous solutions by chemically modified orange peel. J Hazard Mater 185(1):49–54

Fu F, Xie L, Tang B, Wang Q, Jiang S (2012) Application of a novel strategy—advanced Fenton-chemical precipitation to the treatment of strong stability chelated heavy metal containing wastewater. Chem Eng J 189:283–287

García-Gabaldón M, Pérez-Herranz V, García-Antón J, Guinon J (2006) Electrochemical recovery of tin from the activating solutions of the electroless plating of polymers: galvanostatic operation. Sep Purif Technol 51(2):143–149

García-Mendieta A, Olguín MT, Solache-Ríos M (2012) Biosorption properties of green tomato husk (Physalis philadelphica Lam) for iron, manganese and iron–manganese from aqueous systems. Desalination 284:167–174

Gübbük IH, Hatay I, Coşkun A, Ersöz M (2009) Immobilization of oxime derivative on silica gel for the preparation of new adsorbent. J Hazard Mater 172(2):1532–1537

Gulnaziya I, Kheireddine AM, Kim CS (2012) Biomodification of palm shell activated carbon using Aspergillus niger and Bacillus subtilis and its effect on the adsorption of lead ions from aqueous solutions. Afr J Biotechnol 11(82):14812–14821

Gundogdu A, Ozdes D, Duran C, Bulut VN, Soylak M, Senturk HB (2009) Biosorption of Pb(II) ions from aqueous solution by pine bark (Pinus brutia Ten.). Chem Eng J 153(1):62–69

Güzel F, Yakut H, Topal G (2008) Determination of kinetic and equilibrium parameters of the batch adsorption of Mn(II), Co(II), Ni(II) and Cu(II) from aqueous solution by black carrot (Daucus carota L.) residues. J Hazard Mater 153(3):1275–1287

Haron MJ, Tiansih M, Ibrahim NA, Kassim A, Yunus WMZW (2009) Sorption of Cu (II) by poly (Hydroxamic Acid) chelating exchanger prepared from polymethyl acrylate grafted oil palm empty fruit bunch (OPEFB). Bioresources 4(4):1305–1318

Hasan S, Singh K, Prakash O, Talat M, Ho Y (2008) Removal of Cr (VI) from aqueous solutions using agricultural waste 'maize bran'. J Hazard Mater 152(1):356–365

Hashem MA (2007) Adsorption of lead ions from aqueous solution by okra wastes. Int J Phys Sci 2:178–184

Hashem A, Abdel-Halim E, El-Tahlawy KF, Hebeish A (2005) Enhancement of the adsorption of Co (II) and Ni (II) ions onto peanut hulls through esterification using citric acid. Adsorpt Sci Technol 23(5):367–380

Ho Y-S (2003) Removal of copper ions from aqueous solution by tree fern. Water Res 37(10):2323–2330

Ho Y-S, Ofomaja AE (2005) Kinetics and thermodynamics of lead ion sorption on palm kernel fibre from aqueous solution. Process Biochem 40(11):3455–3461

Ho Y-S, Ofomaja AE (2006a) Kinetic studies of copper ion adsorption on palm kernel fibre. J Hazard Mater 137(3):1796–1802

Ho Y-S, Ofomaja AE (2006b) Pseudo-second-order model for lead ion sorption from aqueous solutions onto palm kernel fiber. J Hazard Mater 129(1–3):137–142

Hossain M, Ngo H, Guo W, Nguyen T (2012) Palm oil fruit shells as biosorbent for copper removal from water and wastewater: experiments and sorption models. Bioresour Technol 113:97–101

Ibrahim MNM, Nagah WSW, Norliyana MS, Daud WRW, Rafatullah M, Sulaiman O, Hashim R (2010) A novel agricultural waste adsorbent for the removal of lead (II) ions from aqueous solutions. J Hazard Mater 182(1–3):377–385

Ideriah T, David O, Ogbonna D (2012) Removal of heavy metal ions in aqueous solutions using palm fruit fibre as adsorbent. J Environ Chem Ecotoxicol 4(4):82–90

Igwe J, Abia A (2006) A bioseparation process for removing heavy metals from waste water using biosorbents. Afr J Biotechnol 5(11):1167–1179

Isa MH et al (2008) Removal of chromium (VI) from aqueous solution using treated oil palm fibre. J Hazard Mater 152(2):662–668

Israel U, Eduok U (2012) Biosorption of zinc from aqueous solution using coconut (Cocos nucifera L) coir dust. Arch Appl Sci Res 4(2):809–819

Issabayeva G, Aroua MK, Sulaiman NMN (2006) Removal of lead from aqueous solutions on palm shell activated carbon. Bioresour Technol 97(18):2350–2355

Issabayeva G, Aroua MK, Sulaiman NM (2008) Continuous adsorption of lead ions in a column packed with palm shell activated carbon. J Hazard Mater 155(1–2):109–113

Iyagba ET, Opete OS (2009) Removal of chromium and lead from drill cuttings using activated palm kernel shell and husk. Afr J Environ Sci Technol 3(7):171–179

Ji F, Li C, Tang B, Xu J, Lu G, Liu P (2012) Preparation of cellulose acetate/zeolite composite fiber and its adsorption behavior for heavy metal ions in aqueous solution. Chem Eng J 209:325–333

Kabbashi NA, Elwathig M, Jamil INB (2011) Application of activated carbon from empty fruit bunch (EFB) for mercury [Hg (II)] removal from aqueous solution. Afr J Biotechnol 10(81):18768–18774

Kalinci Y, Hepbasli A, Dincer I (2011) Comparative exergetic performance analysis of hydrogen production from oil palm wastes and some other biomasses. Int J Hydrogen Energy 36(17):11399–11407

Karvelas M, Katsoyiannis A, Samara C (2003) Occurrence and fate of heavy metals in the wastewater treatment process. Chemosphere 53(10):1201–1210

Kazemipour M, Ansari M, Tajrobehkar S, Majdzadeh M, Kermani HR (2008) Removal of lead, cadmium, zinc, and copper from industrial wastewater by carbon developed from walnut, hazelnut, almond, pistachio shell, and apricot stone. J Hazard Mater 150(2):322–327

Khalid N, Ali S, Iqbal A, Pervez S (2007) Sorption potential of styrene-divinylbenzene copolymer beads for the decontamination of lead from aqueous media. Sep Sci Technol 42(1):203–222. doi:10.1080/01496390600957041

Khan MA, Rao RAK, Ajmal M (2008) Heavy metal pollution and its control through non-conventional adsorbents (1998–2007): a review. J Int Environ Appl Sci 3(2):101–141

Khoramzadeh E, Nasernejad B, Halladj R (2012) Mercury biosorption from aqueous solutions by Sugarcane Bagasse. J Taiwan Inst Chem Eng 44(2):266–269

Khraisheh MA, Al-degs YS, Mcminn WA (2004) Remediation of wastewater containing heavy metals using raw and modified diatomite. Chem Eng J 99(2):177–184

Kobya M, Demirbas E, Senturk E, Ince M (2005) Adsorption of heavy metal ions from aqueous solutions by activated carbon prepared from apricot stone. Bioresour Technol 96(13):1518–1521

Ku Y, Chiou H-M (2002) The adsorption of fluoride ion from aqueous solution by activated alumina. Water Air Soil Pollut 133(1–4):349–361

Kurniawan TA, Chan G, Lo W-H, Babel S (2006) Physico–chemical treatment techniques for wastewater laden with heavy metals. Chem Eng J 118(1):83–98

Li Z, Imaizumi S, Katsumi T, Inui T, Tang X, Tang Q (2010) Manganese removal from aqueous solution using a thermally decomposed leaf. J Hazard Mater 177(1–3):501–507

Liang S, Guo X, Tian Q (2011) Adsorption of Pb^{2+} and Zn^{2+} from aqueous solutions by sulfured orange peel. Desalination 275(1):212–216

Low K, Lee C, Tan C (1996) Enhancement of copper sorption through acid blue 29 treated oil palm pressed fibres. Pertanika J Sci Technol 4(1):41–50

Lugo-Lugo V, Barrera-Díaz C, Ureña-Núñez F, Bilyeu B, Linares-Hernández I (2012) Biosorption of Cr (III) and Fe (III) in single and binary systems onto pretreated orange peel. J Environ Manage 112:120–127

Mahmoud ME, Osman MM, Hafez OF, Hegazi AH, Elmelegy E (2010) Removal and preconcentration of lead (II) and other heavy metals from water by alumina adsorbents developed by surface-adsorbed-dithizone. Desalination 251(1):123–130

Malkoc E, Nuhoglu Y (2007) Potential of tea factory waste for chromium (VI) removal from aqueous solutions: thermodynamic and kinetic studies. Sep Purif Technol 54(3):291–298

Mohammad N, Alam MZ, Kabbashi NA, Ahsan A (2012) Effective composting of oil palm industrial waste by filamentous fungi: a review. Resour Conserv Recy 58:69–78

Mohammadi T, Razmi A, Sadrzadeh M (2004) Effect of operating parameters on Pb^{2+} separation from wastewater using electrodialysis. Desalination 167:379–385

Mohammed M, Salmiaton A, Wan Azlina W, Mohammad Amran M, Fakhru'l-Razi A, Taufiq-Yap Y (2011) Hydrogen rich gas from oil palm biomass as a potential source of renewable energy in Malaysia. Renew Sust Energ Rev 15(2):1258–1270

Mohan D, Pittman CU Jr (2006) Activated carbons and low cost adsorbents for remediation of tri- and hexavalent chromium from water. J Hazard Mater 137(2):762–811

Mortaheb HR, Kosuge H, Mokhtarani B, Amini MH, Banihashemi HR (2009) Study on removal of cadmium from wastewater by emulsion liquid membrane. J Hazard Mater 165(1–3): 630–636

Najafi M, Rostamian R, Rafati A (2011) Chemically modified silica gel with thiol group as an adsorbent for retention of some toxic soft metal ions from water and industrial effluent. Chem Eng J 168(1):426–432

Nemr AE (2009) Potential of pomegranate husk carbon for Cr (VI) removal from wastewater: kinetic and isotherm studies. J Hazard Mater 161(1):132–141

Nomanbhay SM, Palanisamy K (2005) Removal of heavy metal from industrial wastewater using chitosan coated oil palm shell charcoal. Electron J Biotechnol 8(1):43–53

Nwabanne JT, Igbokwe PK (2012) Adsorption performance of packed bed column for the removal of lead (ii) using oil palm fibre. Int J Appl Sci Technol 2(5):106–115

Nwabanne JT, Okoye AC, Lebele-Alawa BT (2011) Packed bed column studies for the removal of lead (ii) using oil palm empty fruit bunch. Eur J Sci Res 63(2):296–305

O'Connell DW, Birkinshaw C, O'Dwyer TF (2008) Heavy metal adsorbents prepared from the modification of cellulose: a review. Bioresour Technol 99(15):6709–6724

Ofomaja AE (2010) Equilibrium studies of copper ion adsorption onto palm kernel fibre. J Environ Manage 91(7):1491–1499

Oliveira WE, Franca AS, Oliveira LS, Rocha SD (2008) Untreated coffee husks as biosorbents for the removal of heavy metals from aqueous solutions. J Hazard Mater 152(3):1073–1081

Oluyemi EA, Adeyemi AF, Olabanji IO (2012) Removal of Pb^{2+} and Cd^{2+} ions from wastewaters using palm kernel shell charcoal (PKSC). Res J Eng Appl Sci 1(5):308–313

Onundi YB, Mamun A, Al Khatib M, Ahmed Y (2010) Adsorption of copper, nickel and lead ions from synthetic semiconductor industrial wastewater by palm shell activated carbon. Int J Environ Sci Technol 7(4):751–758

Peng F, Sun R-C (2010) Chapter 7.2—Modification of cereal straws as natural sorbents for removing metal ions from industrial waste water Cereal Straw as a Resource for Sustainable Biomaterials and Biofuels. Elsevier, Amsterdam, pp 219–237

Pereira FV, Gurgel LVA, Gil LF (2010) Removal of Zn^{2+} from aqueous single metal solutions and electroplating wastewater with wood sawdust and sugarcane bagasse modified with EDTA dianhydride (EDTAD). J Hazard Mater 176(1–3):856–863

Prasad AD, Abdullah MA (2009) Biosorption of Fe (II) from aqueous solution using Tamarind Bark and potato peel waste: equilibrium and kinetic studies. J Appl Sci Environ Sanit 4(3):273–282

Qiu H, Lv L, Pan BC, Zhang QJ, Zhang WM, Zhang QX (2009) Critical review in adsorption kinetic models. J Zhejiang Univ Sci A 10(5):716–724

Radzi bin Abas M, Oros DR, Simoneit BR (2004) Biomass burning as the main source of organic aerosol particulate matter in Malaysia during haze episodes. Chemosphere 55(8):1089–1095

Rafatullah M, Sulaiman O, Hashim R, Ahmad A (2010) Adsorption of methylene blue on low-cost adsorbents: a review. J Hazard Mater 177(1):70–80

Rafatullah M, Ahmad T, Ghazali A, Sulaiman O, Danish M, Hashim R (2013) Oil palm biomass as a precursor of activated carbons: a review. Crit Rev Environ Sci Technol 43(11):1117–1161

Rahman M, Awang M, Mohosina B, Kamaruzzaman B, Nik W, Adnan C (2012) Waste palm shell converted to high efficient activated carbon by chemical activation method and its adsorption capacity tested by water filtration. APCBEE Procedia 1:293–298

Rao RA, Rehman F (2010) Adsorption studies on fruits of Gular (Ficus glomerata): removal of Cr(VI) from synthetic wastewater. J Hazard Mater 181(1):405–412

Razmovski R, Šćiban M (2008) Biosorption of Cr (VI) and Cu (II) by waste tea fungal biomass. Ecol Eng 34(2):179–186

Reddy D, Harinath Y, Seshaiah K, Reddy A (2010) Biosorption of Pb(II) from aqueous solutions using chemically modified Moringa oleifera tree leaves. Chem Eng J 162(2):626–634

Reddy D, Ramana D, Seshaiah K, Reddy A (2011) Biosorption of Ni(II) from aqueous phase by Moringa oleifera bark, a low cost biosorbent. Desalination 268(1):150–157

Rupani PF, Singh RP, Ibrahim MH, Esa N (2010) Review of current palm oil mill effluent (POME) treatment methods: vermicomposting as a sustainable practice. World Appl Sci J 11(1):70–81

Saeed A, Akhter MW, Iqbal M (2005) Removal and recovery of heavy metals from aqueous solution using papaya wood as a new biosorbent. Sep Purif Technol 45(1):25–31

Saeed A, Iqbal M, Höll WH (2009) Kinetics, equilibrium and mechanism of Cd^{2+} removal from aqueous solution by mungbean husk. J Hazard Mater 168(2):1467–1475

Saifuddin MN, Kumaran P (2005) Removal of heavy metal from industrial wastewater using chitosan coated oil palm shell charcoal. Electron J Biotechnol 8(1):43–53

Salamatinia B, Zinatizadeh AA, Razali N, Abdullah AZ (2006) Chemical pre-treatments of oil palm frond for improvement in the removal of zn and cu from wastewater by sorption process. Paper presented at the 1st international conference on natural resources engineering and technology 2006, Putrajaya, Malaysia

Salamatinia B, Kamaruddin A, Abdullah A (2007) Removal of Zn and Cu from wastewater by sorption on oil palm tree-derived biomasses. J Appl Sci 7(15):2020–2027

Salleh MAM, Mahmoud DK, Karim WAWA, Idris A (2011) Cationic and anionic dye adsorption by agricultural solid wastes: a comprehensive review. Desalination 280(1):1–13

Sankararamakrishnan N, Sharma AK, Sanghi R (2007) Novel chitosan derivative for the removal of cadmium in the presence of cyanide from electroplating wastewater. J Hazard Mater 148(1–2):353–359

Sekomo CB, Rousseau DP, Saleh SA, Lens PN (2012) Heavy metal removal in duckweed and algae ponds as a polishing step for textile wastewater treatment. Ecol Eng 44:102–110

Sheibani A, Shishehbor MR, Alaei H (2012) Removal of Fe (III) ions from aqueous solution by hazelnut hull as an adsorbent. Int J Ind Chem 3(1):1–3

Silva AM, Cruz FLS, Lima RMF, Teixeira MC, Leão VA (2010) Manganese and limestone interactions during mine water treatment. J Hazard Mater 181(1–3):514–520

Singh TS, Pant K (2004) Equilibrium, kinetics and thermodynamic studies for adsorption of As (III) on activated alumina. Sep Purif Technol 36(2):139–147

Singha B, Das SK (2013) Adsorptive removal of Cu (II) from aqueous solution and industrial effluent using natural/agricultural wastes. Colloids Surf B 1(107):97–106

Srivastava N, Majumder C (2008) Novel biofiltration methods for the treatment of heavy metals from industrial wastewater. J Hazard Mater 151(1):1–8

Sthiannopkao S, Sreesai S (2009) Utilization of pulp and paper industrial wastes to remove heavy metals from metal finishing wastewater. J Environ Manage 90(11):3283–3289

Subbaiah MV, Vijaya Y, Kumar NS, Reddy AS, Krishnaiah A (2009) Biosorption of nickel from aqueous solutions by Acacia leucocephala bark: kinetics and equilibrium studies. Colloids Surf B 74(1):260–265

Sugawara K, Wajima T, Kato T, Sugawara T (2007) Preparation of carbonaceous heavy metal adsorbent from palm shell using sulfur impregnation. Ars Separatoria Acta 5:88–98

Sulaiman O, Salim N, Hashim R, Yusof LHM, Razak W, Yunus NYM, Hashim WS, Azmy MH (2009) Evaluation on the suitability of some adhesives for laminated veneer lumber from oil palm trunks. Mater Des 30(9):3572–3580

Sulaiman O, Amini M, Hazim M, Rafatullah M, Hashim R, Ahmad A (2010) Adsorption equilibrium and thermodynamic studies of copper (II) ions from aqueous solutions by oil palm leaves. Int J Chem React Eng 8(1):1–18

Sun Y, Webley PA (2010) Preparation of activated carbons from corncob with large specific surface area by a variety of chemical activators and their application in gas storage. Chem Eng J 162(3):883–892

Tan W, Ooi S, Lee C (1993) Removal of chromium (VI) from solution by coconut husk and palm pressed fibres. Environ Technol 14(3):277–282

Tan W, Lee C, Ng K (1996) Column studies of copper (II) and nickel (II) ions sorption on palm pressed fibres. Environ Technol 17(6):621–628

Tan I, Ahmad A, Hameed B (2008) Optimization of preparation conditions for activated carbons from coconut husk using response surface methodology. Chem Eng J 137(3):462–470

Tijani JO (2011) Sorption of lead (ii) and copper (ii) ions from aqueous solution by acid modified and unmodified Gmelina Arborea (Verbenaceae) leaves. J Emerg Trend Eng Appl Sci 2(5):734–740

Torab-Mostaedi M, Asadollahzadeh M, Hemmati A, Khosravi A (2013) Equilibrium, kinetic, and thermodynamic studies for biosorption of cadmium and nickel on grapefruit peel. J Taiwan Inst Chem Eng 44(2):295–302

Tsekova K, Todorova D, Ganeva S (2010) Removal of heavy metals from industrial wastewater by free and immobilized cells of Aspergillus niger. Int Biodeter Biodegr 64(6):447–451

Uemura Y, Omar WN, Tsutsui T, Yusup SB (2011) Torrefaction of oil palm wastes. Fuel 90(8):2585–2591

Urgun-Demirtas M, Benda PL, Gillenwater PS, Negri MC, Xiong H, Snyder SW (2012) Achieving very low mercury levels in refinery wastewater by membrane filtration. J Hazard Mater 215:98–107

Vaghetti JC et al (2009) Pecan nutshell as biosorbent to remove Cu (II), Mn (II) and Pb (II) from aqueous solutions. J Hazard Mater 162(1):270–280

Vargas AM, Garcia CA, Reis EM, Lenzi E, Costa WF, Almeida VC (2010) NaOH-activated carbon from flamboyant (Delonix regia) pods: optimization of preparation conditions using central composite rotatable design. Chem Eng J 162(1):43–50

Vázquez G, Calvo M, Sonia Freire M, González-Alvarez J, Antorrena G (2009) Chestnut shell as heavy metal adsorbent: optimization study of lead, copper and zinc cations removal. J Hazard Mater 172(2):1402–1414

Venugopal V, Mohanty K (2011) Biosorptive uptake of Cr(VI) from aqueous solutions by Parthenium hysterophorus weed: equilibrium, kinetics and thermodynamic studies. Chem Eng J 174(1):151–158

Vieira MGA, Neto AFA, Gimenes ML, da Silva MGC (2010) Sorption kinetics and equilibrium for the removal of nickel ions from aqueous phase on calcined Bofe bentonite clay. J Hazard Mater 177(1–3):362–371

Vinod V et al (2011) Bioremediation of mercury (II) from aqueous solution by gum karaya (Sterculia urens): a natural hydrocolloid. Desalination 272(1):270–277

Wahi R, Ngaini Z, Jok VU (2009) Removal of mercury, lead and copper from aqueous solution by activated carbon of palm oil empty fruit bunch. World Appl Sci J 5:84–91

Wan Nik W, Rahman M, Yusof A, Ani F, Adnan C (2006) Production of activated carbon from palm oil shell waste and its adsorption characteristics. In: 1st international conference on natural resources engineering and technology 2006, Putrajaya, Malaysia, pp 646–654

Wang S, Peng Y (2010) Natural zeolites as effective adsorbents in water and wastewater treatment. Chem Eng J 156(1):11–24

Wang X-S, Qin Y (2006) Removal of Ni(II), Zn(II) and Cr(VI) from aqueous solution by Alternanthera philoxeroides biomass. J Hazard Mater 138(3):582–588

Wang L, Wang R, Oliveira R (2009) A review on adsorption working pairs for refrigeration. Renew Sust Energ Rev 13(3):518–534

Xing Y, Chen X, Wang D (2007) Electrically regenerated ion exchange for removal and recovery of Cr (VI) from wastewater. Environ Sci Technol 41(4):1439–1443

Yadla SV, Sridevi V, Lakshmi MVVC (2012) A review on adsorption of heavy metals from aqueous solution. J Chem Biol Phys Sci 2(3):585–1593

Yin C, Aroua M, Daud W (2008a) Enhanced adsorption of metal ions onto polyethyleneimine-impregnated palm shell activated carbon: equilibrium studies. Water Air Soil Pollut 192 (1–4):337–348. doi:10.1007/s11270-008-9660-9

Yin CY, Aroua MK, Daud WMAW (2008b) Polyethyleneimine impregnation on activated carbon: effects of impregnation amount and molecular number on textural characteristics and metal adsorption capacities. Mater Chem Phys 112(2):417–422

Ying X, Fang Z (2006) Experimental research on heavy metal wastewater treatment with dipropyl dithiophosphate. J Hazard Mater 137(3):1636–1642

Zahir F, Rizwi SJ, Haq SK, Khan RH (2005) Low dose mercury toxicity and human health. Environ Toxicol Pharmacol 20(2):351–360

Environmental Fate and Toxicology of Chlorothalonil

April R. Van Scoy and Ronald S. Tjeerdema

Contents

1 Introduction 89
2 Chemistry 90
3 Chemodynamics 91
   3.1 Soil 91
   3.2 Water 92
   3.3 Air 93
4 Environmental Degradation 93
   4.1 Abiotic Processes 93
   4.2 Biotic Processes 96
5 Toxicology 97
   5.1 Mode of Action 97
   5.2 Aquatic Organisms 99
   5.3 Mammals 100
   5.4 Birds 100
   5.5 Plants 101
   5.6 Fungi 101
6 Summary 102
References 103

1 Introduction

The fungicide chlorothalonil (2,4,5,6-tetrachloro-1,3-benzenedicarbonitrile; CAS 1897-45-6; Fig. 1) was introduced in 1965 by Diamond Shamrock Corp. and was first registered in 1966 for use on turfgrass within the United States. An additional registration was granted 4 years later for use on potatoes, marking it the first approved food crop for application (US EPA 1999). It is formulated as concentrates,

---

A.R. Van Scoy (✉) • R.S. Tjeerdema
Department of Environmental Toxicology, College of Agricultural & Environmental Sciences, University of California, One Shields Ave, Davis, CA 95616-8588, USA
e-mail: avanscoy@ucdavis.edu

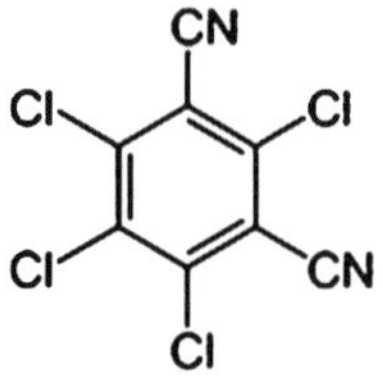

Fig. 1 Chlorothalonil structure

powders, and granules, among other registered formulations. Some of the prominent products containing chlorothalonil as the active ingredient include Bravo®, Daconil® and Sweep® (US EPA 1999). These or other chlorothalonil formulations have been applied to crops such as celery, beans, peanuts, and peaches, among others. Within the USA, approximately 34% of the total chlorothalonil applied is used on peanuts, 12% on potatoes and 10% on golf courses (US EPA 1999).

Chlorothalonil is a broad spectrum, non-systemic, organochlorine fungicide and mildewicide It is principally used to control fungal foliar diseases on various fruits, vegetables, ornamentals and turf (US EPA 1999). Chlorothalonil's success as an antifouling paint additive and wood protectant qualified it to replace organotin biocides such as tributyltin; however, once applied, it is slowly released into waterways and potentially contaminates surface water bodies (Sakkas et al. 2002). Although surface waters near marinas in San Diego, CA were monitored for such antifouling residues, none were detected above a detection limit of 10 ng/L (Sapozhnikova et al. 2007). In California, surface and groundwater were monitored for chlorothalonil residues from 1993 to 2000. Of the samples collected (705 total) from USGS water monitoring stations, only one surface water sample contained chlorothalonil at a concentration of 0.29 µg/L (USGS NAWQA; US EPA 2007).

Chlorothalonil has a low water solubility and is moderately persistent in soils, having half-lives ($t_{1/2}$s) up to 19 days. Because of its water solubility, the potential for chlorothalonil to impact groundwater is low; however, it has been found to highly impact aquatic organisms (US EPA 1999). The environmental fate of chlorothalonil was last reviewed in the mid 1990s (Caux et al. 1996). The goal of this paper is to review the relevant literature that has appeared since 1996, focusing on chlorothalonil's chemistry, environmental fate and toxicity.

2 Chemistry

Chlorothalonil is a chloronitrile fungicide (Tomlin 2000), and specifically is a polychlorinated aromatic (US EPA 1999). Technical grade chlorothalonil is an odorless or slightly pungent, colorless crystalline solid. Chlorothalonil is insoluble in water (at 25 °C), but is slightly soluble in kerosene, acetone and xylene, and this compound strongly adsorbs to soil and sediment. Chlorothalonil is denser than water, potentially susceptible to hydrolysis under alkaline conditions, stable against photolysis and is degraded by both aerobic and anaerobic microbes. Additional physiochemical properties of chlorothalonil are presented in Table 1.

Table 1 Physiochemical properties of chlorothalonil

Chemical Abstracts Service registry number (CAS#)[a]	1897-45-6
Molecular Formula[a]	$C_8Cl_4N_2$
Molecular weight (g/mol)[a]	265.9
Density at 20 °C (g/mL)[a]	2.0
Melting point (°C)[a]	252.1
Octanol-water partition coefficient (log K_{ow})[b]	2.88
Organic carbon normalized partition coefficient (K_{oc})[c]	5,000
Vapor pressure at 25 °C (torr)[b]	5.72×10^{-7}
Henry's law constant atm m^{-3} mol[b]	1.4×10^{-7}
Solubility (g/kg)[a]	0.81
Water (mg/L)	<10
Kerosene	20
Acetone	80
Xylene	

[a]Data from Tomlin (2000)
[b]CA DPR Risk Characterization Document (2005)
[c]Data from Waltz et al. (2002)

3 Chemodynamics

3.1 Soil

Chlorothalonil has the potential to strongly adsorb to soil and sediment, as indicated by its high K_{oc} constant. Adsorption isotherms on five clay minerals (montmorillonite, Na-bentonite, Ca-bentonite, allophone and kaolinate) and three soils, having an organic carbon content of 1.1, 1.4 and 5.2%, respectively, were investigated by Fushiwaki and Urano (2001). Based on the Freundlich isotherm equation, chlorothalonil had a lower adsorption capacity (k_f values ranged from 70 to 2,000) than did pentachlorothioanisole ($k_{f\,values}$ ranged from 4,400 to 30,000). In addition, n-values ranged from 1.3 to 1.8 for each of the soils and clays (Fushiwaki and Urano 2001). Furthermore, the adsorption rate was not linked to organic carbon content; however, it may be influenced by inorganic matter.

Patakioutas and Albanis (2002) investigated the trend between adsorption and organic matter (OM) content. Soils of varying OM content and varying concentrations of chlorothalonil (0.1–0.5 mg/L) produced three adsorption isotherm shapes. As soil OM content increased, the shape of the isotherm changed from S- to L- to C-shape and k_f values respectively ranged from 96.3 to 1,356.9 (Patakioutas and Albanis 2002). The results of this study illustrated the strength of OM in immobilizing pesticides.

Chlorothalonil adsorbs most strongly to soils that have high organic matter, silt and clay. It has a low affinity to bind to sand, thus it is moderately to highly mobile in sandy soils (US EPA 1999). To investigate this, Gamble et al. (2000), analyzed the distribution of chlorothalonil among a quartz sand soil. The soil (Simcoe: 90–95% quartz sand) was placed in solution microcosms. After 14 days, 43.3% of the chlorothalonil remained in solution, 26.2% resided in the labile sorbed state and 30.5% existed as a bound residue. It is thought that the 5–10% non-quartz material was responsible for sorbing the measured bound residues (Gamble et al. 2000).

The half-life of chlorothalonil that had been applied to a low-humic sandy soil was 12 days; 45% of the parent compound had been transformed into one major metabolite hydroxychlorothalonil (van der Pas et al. 1999). Furthermore, movement of this metabolite through the soil was decreased from adsorption, although low concentrations were measured in groundwater (van der Pas et al. 1999). Wang et al. (2009) determined the half-lives for chlorthalonil on both non-sterilized and sterilized non-amended soil (containing sandy loam, sand and clay) to be 8.8 and 19 days, respectively.

To address the possibility of soil runoff, Potter et al. (2001) investigated degradation rates and soil surface residues from peanut plots (Tifton loamy sand) treated with seven successive chlorothalonil applications (1.25 kg/ha; 2-week intervals). Soil residues were highest following the second application, however concentrations decreased as plant canopies obstructed disposition. Half-lives were determined for both chlorothalonil ($t_{1/2}$ = <1–3.5 days) and its primary product 4-hydroxychlorothalonil ($t_{1/2}$ = 10–22 days); further breakdown products had half-lives 10–20 times longer than chlorothalonil (Potter et al. 2001). Waltz et al. (2002) also confirmed that the known metabolite hydroxychlorothalonil (HC) is more persistent in soil compared to its parent. In summary, chlorothalonil is regarded to remain bound to soil, primarily because it has low water solubility and a high Koc constant.

3.2 Water

Pesticides that are used on turf grasses and other vegetation pose a potential risk of leaching into groundwater. Wu et al. (2002) evaluated chlorthalonil's potential to leach and the distance it travels in soil. Because of its low water solubility, chlorothalonil displayed a negligible tendency to leach in soil, as evidenced by its retention in the upper 0.2 cm thatch layer in soil samples collected before, and 0, 2, 7, 15, 30, 61, 83, and 120 days following treatment (Wu et al. 2002). Armbrust (2001) measured leachate for both chlorothalonil and its degradate, hydroxylchlorothalonil (HC). Of 130 samples analyzed, HC was found in 87% of the samples, but chlorothalonil was detected in only one. Although HC is persistent in soil, the evidence indicates that it is rapidly photodegraded under aqueous conditions and has a half-life of 35 min; hence, HC should not pose a potential risk to surface water (Armbrust 2001).

The potential for chlorothalonil to run off of application sites was simulated by Haith and Rossi (2003). Mean annual runoff concentrations for golf course greens, in three U.S. cities (Boston, Philadelphia and Rochester) were determined to be 0.477, 0.699 and 0.372 mg/L, respectively, whereas for fairways, concentrations were 0.296, 0.450 and 0.256 mg/L, respectively (Haith and Rossi 2003). For both greens and fairways, these measured concentrations exceeded the aquatic 96-h LC_{50} values for both the rainbow trout and water flea. The use of chlorothalonil on peanut fields, particularly in U.S. regions that have increased rainfall, increases the potential to contaminate local streams and ponds. However, the presence of increased plant foliage may decrease leaching of this chemical, although the degree of foliar wash-off for chlorothalonil has not been determined (Potter et al. 2001).

3.3 Air

The rate of volatilization of chlorothalonil from water, dry and moist soil is low, as predicted by its having low vapor pressure and Henry's law constant values (Table 1). Because of the low vapor pressure, initial volatilization is slow and volatility loss continues over a longer time period (Leistra and Van den Berg 2007). In general, the volatilization of chlorothalonil can be regarded as negligible and does not represent a significant dissipation route.

Bedos et al. (2010) measured chlorothalonil levels in air shortly after the fungicide was applied to wheat (theoretical application dose of 880 g/ha; application volume of 150 L/ha) in May of 2006. Measurable air concentrations were recorded in the human breathing zone (0.68 m above the soil) following the application. A cumulated volatilization flux, after 31 h, was determined to be 5 g/ha, respectively, which represents an approx. loss of 0.6% of the theoretical application dose. Air concentrations decreased slightly over 6 days (from 28 $\mu g/m^3$ to 64 ng/m^3), and a volatilization flux of 17.5 g/ha was estimated for this compound (Bedos et al. 2010).

4 Environmental Degradation

4.1 Abiotic Processes

4.1.1 Hydrolysis

Chlorothalonil is stable to hydrolysis at pH 5 and 7 (Szalkowski and Stallard 1977; US EPA 1999). However, under basic conditions (pH 9), the compound degrades to form two products: 3-cyano-2,4,5,6-tetrachlorobenzamide and 4-hydroxyl-2,5,6-trichloroisophthalonitrile (Szalkowski and Stallard 1977). Kwon and Armbrust (2006) proposed that the pathway for chlorothalonil degradation in aquatic systems would proceed by reductive dechlorination, oxidative dechlorination/hydrolysis and base hydrolysis (Fig. 2). The US EPA (1999) reported a hydrolysis half-life value for chlorothalonil of 30–60 days.

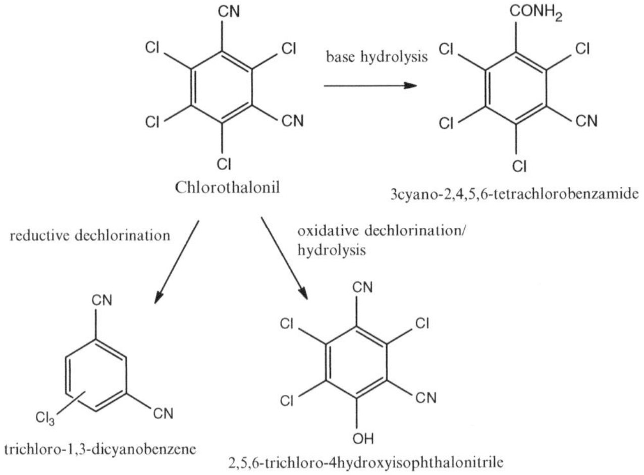

Fig. 2 Proposed aquatic degradation pathway for chlorothalonil (Adapted from Kwon and Armbrust 2006)

4.1.2 Photolysis

Aqueous dissolved concentrations of chlorothalonil absorb sunlight within the wavelength range of 300–340 nm, and direct photolysis represents a major degradation pathway for this fungicide (Leistra and Van Den Berg 2007). Chlorothalonil, exposed directly to light (300–400 nm) photolytically degraded more rapidly in natural waters ($DT_{50}=0.21$–0.76 days) than in a buffered aqueous system (pH 7; $DT_{50}=1.1$ days; Wallace et al. 2010). Monadjemi et al. (2011) investigated the photodegradation of chlorothalonil on a simulated plant surface, specifically using paraffin wax (irradiated at wavelengths between 300 and 800 nm). A field-extrapolated half-life of 5.3 days resulted, and suggested that chlorothalonil is susceptible to direct photolysis, in addition to surface penetration. In addition, these authors found the main degradation route was via reductive dechlorination (Monadjemi et al. 2011). Waltz et al. (2002) studied the photodegradation of hydroxychlorothalonil (HC), chlorothalonil's major hydrolytic metabolite. Results were that HC in the water samples exposed to simulated sunlight (via use of lamps) absorbed radiation, and this substance was photolyzed with a $t_{1/2}$ of 33–37 min).

Degradation of chlorothalonil via the Fenton reaction (Fe^{3+}/H_2O_2; Fig. 3) was effectively achieved by Park et al. (2002). Half-lives were determined under dark ($t_{1/2}=77$ min) and UV irradiated conditions ($t_{1/2}=49.5$ min), and results indicate that breakdown was enhanced by increased ferric ion concentrations (dark $t_{1/2}=31.7$ min and UV $t_{1/2}=16.9$ min). This reaction proceeds by dechlorination of chlorothalonil.

Penuela and Barcelo (1998) investigated the influence of water quality and photosensitizers (TiO_2 and $FeCl_3$) on the degradation of chlorothalonil, by using a xenon arc lamp and natural sunlight. They found that the $t_{1/2}$ of chlorothalonil in deionized

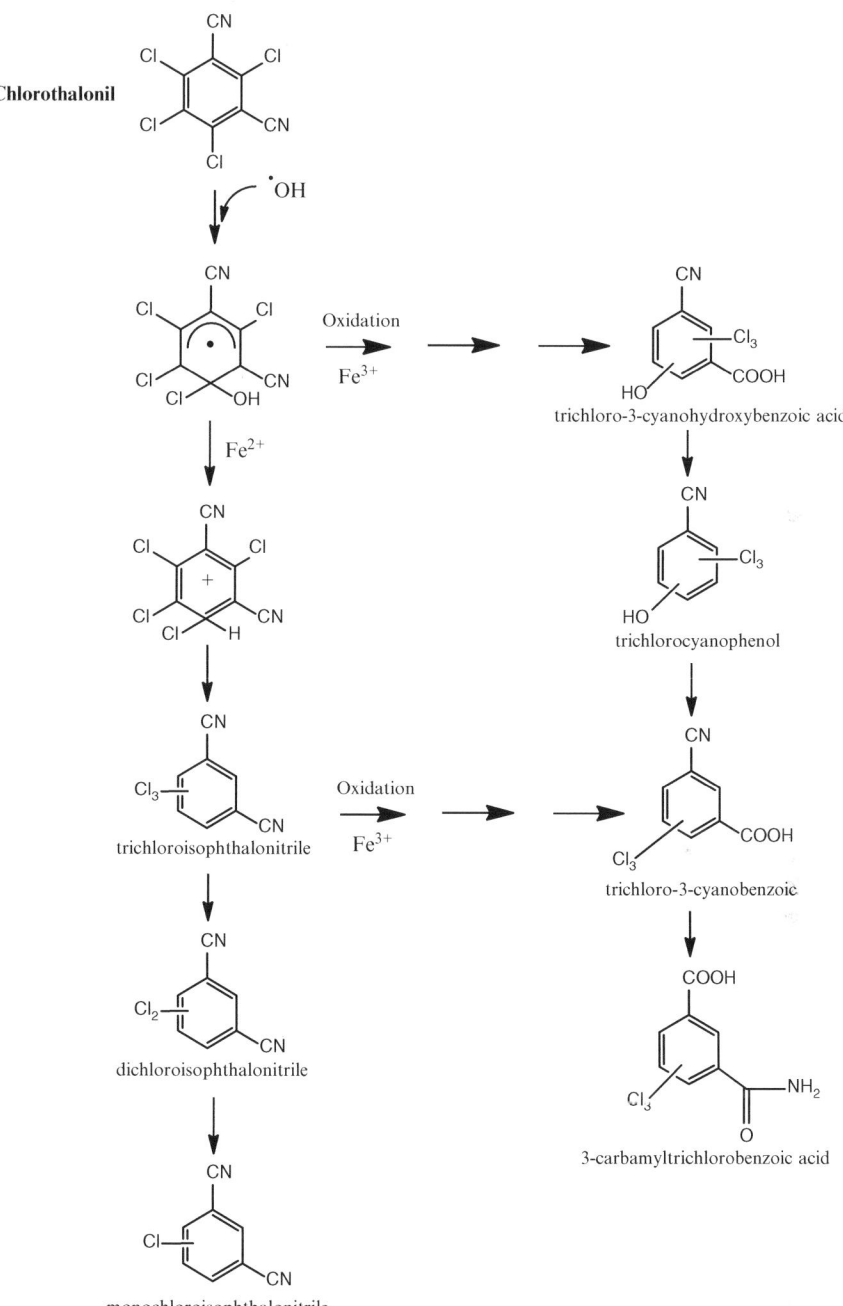

Fig. 3 Proposed breakdown pathway of chlorothalonil treated with Fenton reagent (Fe^{3+}/H_2O_2). (Adapted from Park et al. 2002)

water (101.17 h; sunlight) was longer than deionized water irradiated under a xenon lamp (36.86 h); groundwater irradiated with the lamp had a half-life of 0.71 h. Moreover, addition of the photosensitizers decreased half-lives as well; $FeCl_3$ was a better catalyst under lamp irradiated water ($t_{1/2}$ = 1.37 h) than water irradiated by sunlight ($t_{1/2}$ = 4.24 h). The results of this study demonstrated that degradation follows first-order kinetics in the presence of catalysts (Penuela and Barcelo 1998).

Studies by Sakkas et al. (2002) also showed that photolysis of chlorothalonil follows pseudo-first order kinetics. The photolytic degradation in waters from a river and a lake was determined to be more rapid (99% loss within 60 h) than in distilled or seawater (67 and 72% loss, respectively), when irradiated under both natural and simulated conditions. Major photoproducts (viz., chloro-1,3-dichlorobenzene, dichloro-1,3-dicyanobenzene, trichloro-1,3-dicyanobenzene and benzamide) were identified in this study (Sakkas et al. 2002). It is thought that the presence of dissolved organic matter (DOM) and other photosensitizers may have enhanced the rate of photodegradation. To investigation, studies which included photosensitizers indicated that increased concentrations of bicarbonate promoted degradation rates, whereas, degradation via carbonate radicals ($\cdot CO_3^-$) dominated under situations, in which degradation via the hydroxyl radical ($\cdot OH$) was minimal (Wallace et al. 2010). In summary, direct photolysis of chlorothalonil proceeds rapidly and is enhanced by the presence of photosensitizers.

4.2 Biotic Processes

Microbial digestion is thought to be the primary pathway by which chlorothalonil is degraded (US EPA 1999). Chen et al. (2001) studied the effects of microbes on fungicides in three soil types (Canfield silt-loam Luvisol; pH 6.3; unamended and amended with alfalfa leaves and wheat straw). They found chlorothalonil inhibits microbial activity in the treated soils. In unamended soil, enhanced mineralization and decreased nitrification rates occurred. Mori et al. (1996) evaluated the microbial degradation of chlorothalonil in unfertilized and fertilized (farmyard manure) soil. Microbial activity was enhanced in soil treated with a combination of chemical and farmyard fertilizer and degradation increased as soil pH reached neutrality. Incorporating manure in the soil stimulated the microbes, although they required additional carbon sources (Mori et al. 1996). Wang et al. (2011) studied chlorthalonil's anaerobic degradation in four paddy soils. In these studies, soil pH and total carbon content both highly affected the rate of biodegradation. Results indicate that chlorothalonil was more efficiently degraded under neutral pH (6.3–6.6) conditions and in soil containing 3–4% total carbon (Wang et al. 2011).

Motonaga et al. (1996) identified the gram-negative rod bacterium, TB 1, from chlorothalonil-treated soil. This bacterium transformed more than 75% of chlorothalonil present in the soil into 4-hydroxy-2,5,6-trichloroisophthalonitrile and chloride anion via hydrolysis, rather than via mineralization. Out of 50 identified chlorothalonil degrading bacteria, the TB 1 strain was the only one to produce the

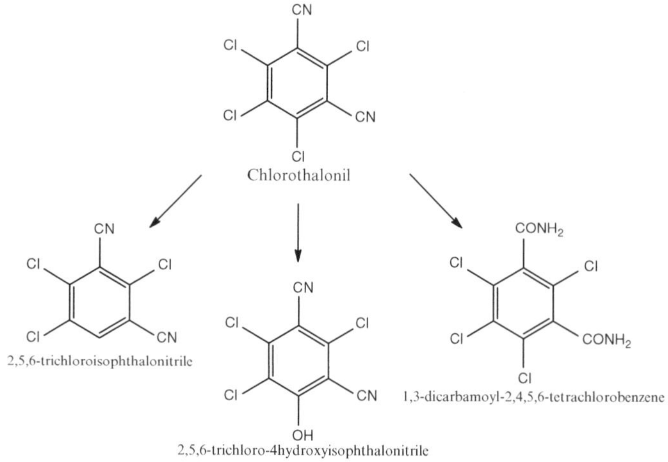

Fig. 4 Proposed microbial degradation pathways for chlorothalonil via dechlorination and hydroxylation (Adapted from Ukai et al. 2003)

hydroxylated metabolite (Motonaga et al. 1996). Zhang et al. (2007) observed the NS1 strain of *Bacillus cereus* to degrade chlorothalonil as a result of cometabolism, and carbon sources enhanced its degradation. Liang et al. (2010) isolated the bacterial strain CTN-11 (identified as an *Ochrobactrum* sp.) from chlorothalonil-contaminated soil. This strain degraded chlorothalonil to undetectable levels within 48 h when exposed to a temperature range of 20–40 °C and a pH from 6 to 9. Under anaerobic conditions, hydrolytic dechlorination occurred, producing the more stable hydroxy metabolite (Liang et al. 2010).

The influence of the chlorothalonil chlorine atoms on degradation was examined by Ukai et al. (2003). They found that chlorothalonil degradates appear to contain 3–4 chlorine atoms, and these degradates suppress soil degradation of the parent compound. The two major degradate products (Fig. 4) were 2,5,6-trichloro-4-hydroxyisophthalonitrile and 2,5,6-trichloroisophthalonitrile. Other degradation products were identified by Sato and Tanaka (1987). They also concluded that degradation occurred via dechlorination and partial substitution (Fig. 5). The possible degradation products for chlorthalonil are listed in Table 2.

5 Toxicology

5.1 Mode of Action

The fungicidal activity exhibited by chlorothalonil is attributed to the inactivation of cell sulfhydryl enzymes (Vincent and Sisler 1968; Sherrard et al. 2003). Gallagher et al. (1992) recorded a depletion of glutathione, resulting in the inhibition of

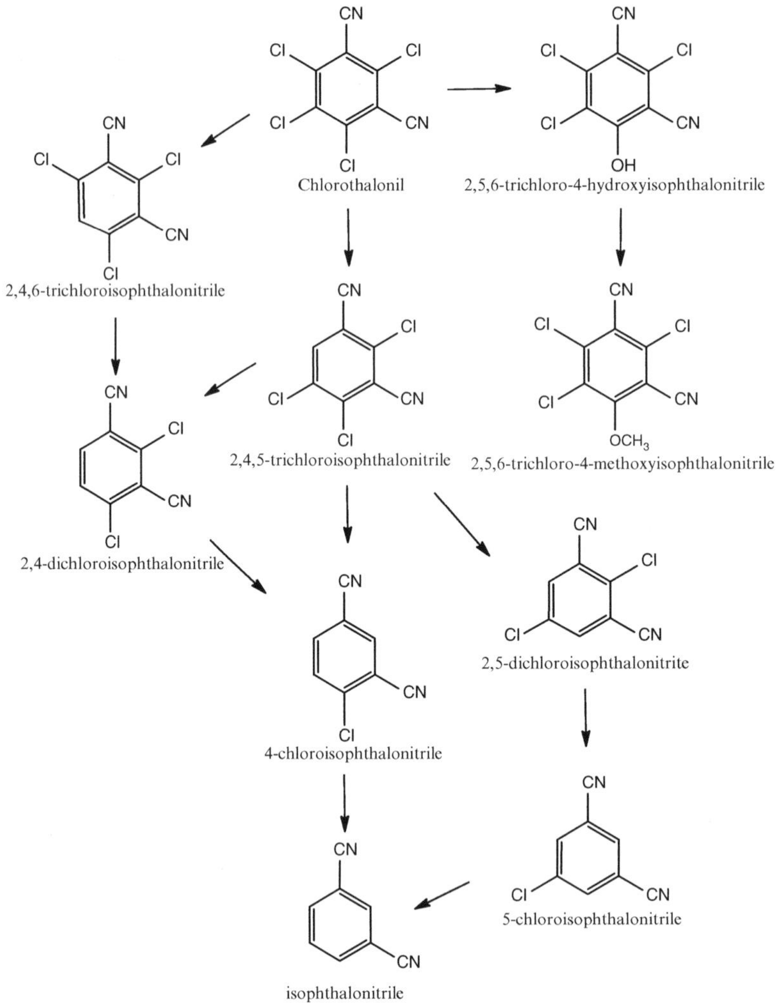

Fig. 5 Proposed soil degradation pathway for chlorothalonil (Adapted from Sato and Tanaka 1987)

Table 2 Possible microbial degradation products[a]

Compound name	Soil conditions	Metabolite
4-hydroxychlorothalonil	Aerobic	Major
Metylthiotrichloroisophtalonitrile	Aerobic	Major
3-carbamyl-2,4,5-trichlorobenzoic acid	Aerobic acidic	Major
3-cyano-2,3,4,5,6-tetrachlorobenzoamide	Aerobic acidic	Major
Trichloroisophtalonitrile	Aerobic	Minor
m-phthalonitrile	NA	Breakdown product

[a]Data from Carlo-Rojas et al. (2004)

Table 3 Toxicity (expressed as 48-h or 96-h LC_{50} values) of technical grade chlorothalonil to aquatic organisms

Aquatic organism	Scientific name	Concentration (µg/L)
Rainbow trout[a]	Lepomis macrochirus	10.5–76
Fathead minnow[a]	Pimephales promelas	23
Bluegill[a]	Lepomis macrochirus	51–84
Waterflea[a]	Daphnia magna	54–68
Pink Shrimp[b]	Penaeus duorarum	154

[a]Data from US EPA (2007)
[b]Data from US EPA (1999)

glucose oxidation in exposed channel catfish. A study, in which *Saccharomyces pastorianus* and *Neurospora crassa* were exposed to chlorothalonil confirmed that glucose oxidation was impaired and soluble thiol content was reduced from chlorothalonil treatment (Vincent and Sisler 1968). Tillman et al. (1973) concluded that the mechanism of chlorothalonil's toxic action resembles that of the trichoromethylsulfenyl fungicides. Although many studies have examined chlorothalonil's mode of action, chlorothalonil and other chloronitriles have been categorized as having multiple sites of action; resistance to the fungicide does not develop (FRAC 2013).

5.2 Aquatic Organisms

The potential for chlorothalonil to bioaccumulate in aquatic species is relatively low because it aggressively binds to soils. Yet, exposure from sediment-bound residues is possible. Although it is assumed that bioaccumulation will be minimal, chlorothalonil has been found to be highly toxic to many aquatic species. For example, it is highly toxic to fathead minnow (*Pimephales promelas*) and somewhat less toxic to *Daphnia magna* and pink shrimp (*Penaeus duorarum*; Table 3).

Early life-stages of the freshwater mussel, *L. siliquoidea*, were exposed to selected technical-grade pesticides. Chlorothalonil was more toxic to glochidia (48-h $EC_{50}=0.04$ mg/L) than to juvenile mussels (96-h $EC_{50}=0.28$ mg/L), and had higher toxicity than other pesticides such as atrazine and fipronil (Bringolf et al. 2007).

Larval and adult stages of the grass shrimp, *Palaemonetes pugio*, were exposed to a range of chlorothalonil concentrations, and thereafter exhibited increased toxicity with increasing temperature (25° vs. 35 °C) and salinity (20 ppt vs. 30 ppt; DeLorenzo et al. 2009). Furthermore, 96-h LC_{50} values for larvae were more variable among exposure conditions. Under standard and high salinity conditions 96-h LC_{50} values were 49.1 and 39.4 µg/L, respectively. In addition, 96-h LC_{50}s for adult shrimp were 156 and 116 µg/L, respectively, under the same conditions. Generally, results show that toxicity increased as exposure length increased from 24 to 96 h (DeLorenzo et al. 2009).

Laboratory and field bioassays were conducted to determine the potential hazard chlorothalonil poses towards aquatic fauna. Rainbow trout (96-h $LC_{50}=69$ μg/L) was more sensitive than blue mussels (96-h $LC_{50}=5.94$ mg/L) and the water flea (48-h $EC_{50}=97$ μg/L), when exposed under laboratory conditions (Ernst et al. 1991). However, caged organisms, exposed under field conditions (aerially treated pond), were less sensitive, and exposed rainbow trout did not suffer any mortality. Ernst et al. (1991) concluded that environmental factors such as, microbial degradation, dilution and adsorption to suspended matter reduced chlorothalonil's toxicity.

The toxicity and site of chlorothalonil accumulation was investigated by Davies and White (1985). Four fish species (*Salmo gairdneri, Galaxias maculates, G. truttaceus and G. auratus*) were exposed under flow-through conditions (≤ 0.6 mg/L, 13–16 °C, $[O_2]=8$ mg/L), and exhibited increased toxicity; 96-h LC_{50} values ranged from 16.3 to 29.2 μg/L. In addition, using radiotracers (10 μg/L; 96-h) Davies and White (1985) found that ^{14}C-CN labelled chlorothalonil to be highly accumulated within the gall bladder and hind gut of each species.

5.3 Mammals

Groups of pregnant female mice were orally administered chlorothalonil at doses ranging from 0 to 600 mg/kg/day. Although the treatments produced no mortality, signs of toxicity, such as weakness and reduced activity did occur at the 400 and 600 mg/kg/day dose levels (Farag et al. 2006). Also observed at these concentrations was significant embryo lethality and a reduction in live fetuses (Farag et al. 2006). According to the US EPA (1999), chlorothalonil is considered to be practically non-toxic to small mammals, based on having a measured rat LD_{50} of >10,000 mg/kg. However, a known degradate, SDS-3701 is much more acutely toxic than the parent compound (viz., possessing an acute female rat LD_{50} of 242 mg/kg). This degradate possesses high chronic oral toxicity towards pregnant rabbits and has a developmental no observable effect level (NOEL) of 33 mg/L, compared to chlorothalonil itself (NOEL=330 mg/L; US EPA 1999).

Mozzachio et al. (2008) investigated the incidence of pesticide applicators that were both exposed to chlorothalonil and were diagnosed with cancer. They found no direct link to applicators with colon, lung or prostate cancers; approximately 3,600 applicators used chlorothalonil an average of 3.5 days per year. Although animal studies have provided sufficient evidence to classify chlorthalonil as a probable carcinogen, it is not known if it is a human carcinogen or not (Mozzachio et al. 2008).

5.4 Birds

Chlorothalonil is acutely non-toxic to birds when administered orally; LD_{50} values range from >2,000 mg/kg-bwt for Japanese quail to >10,000 mg/kg-bwt for both mallard and northern bobwhite quail (US EPA 2007). Reproductive effects caused by dietary exposure have been investigated in bobwhite quail. At the highest dose of 10,000 mg/kg, reproductive impairment occurred and caused effects on general health and hatching survival. Additional studies with Mallard ducks were conducted and decreased egg production was observed (WHO 1996). Although chlorothalonil's toxicity is low to birds, similar to what occurs in mammals, its degradate SDS-3701 is much more toxic. Avian studies have shown that Mallard ducks are the most sensitive bird species to the toxicity of SDS-3701, which has an acute LD_{50} of 158 mg/kg (US EPA 2007).

5.5 Plants

Chlorothalonil residues that appear on foliar surfaces after application to various crops have been investigated. Putman et al. (2003) used cranberries to evaluate dislodgeable foliar and fruit residues following application of chlorothalonil with and without an adjuvant. Two applications were made: one at 20% cranberry blossom bloom and another at 80% bloom (14 days later). Measured dislodgeable foliar residue concentrations were found to increase with the use of an adjuvant; the estimated half-life for chlorothalonil with and without adjuvant was determined to be 12 and 13 days, respectively. Furthermore, the cranberries were harvested 76 days post-application, and showed fruit residues of chlorothalonil and its metabolites 4-hydroxy-2,5,6-trichloroisophthalonitrile and 1,3-dicarbamoyl-2,4,5,6-tetrachlorobenzene (Putnam et al. 2003).

Not only is chlorothalonil present on foliar surfaces, but it can also cause oxidative stress if taken up by plants. An experiment on upland rice (*Oryza sativa*) was conducted to determine the impact of chlorothalonil application on the plant with or without the presence of arbuscular mycorrhizal fungus (AMF; *Glomus mosseae*). Under both conditions, plant growth was significantly inhibited and the presence of fungi decreased phosphorous concentrations within plant shoots (Zhang et al. 2006). Further investigation showed chlorothalonil to induce oxidative stress, and affect catalase, ascorbic peroxidase, and peroxidase activity (Zhang et al. 2006).

5.6 Fungi

The effectiveness of chlorothalonil as a fungicide has been studied on vesicular arbuscular mycorrhizal (VAM) *Glomus aggregatum* fungi. Chlorothalonil was mixed into Wahiawa silty clay soil (at concentrations ranging from 0 to 200 mg ai/kg soil), and the applied levels decreased VAM colonization with increasing concentrations (Habte et al. 1992). In addition, Habte et al. (1992) noted that chlorothalonil toxicity persisted for 12.5 weeks after initial soil application. Exposure of the VAM *G. intraradices* fungi, at 0.13 mg/L, reduced overall VAM formation

(Wan et al. 1998). They also found the concentration at which growth and development was inhibited by 50% to be 0.05 ± 0.01 mg/L for extraradical mycelial growth and 0.04 ± 0.009 mg/L, respectively, following a 14 days inoculation.

Latteur and Jansen (2002) investigated the ability of 20 fungicides to affect the infectivity of conidia of the fungus *E. neoaphidis*- an insect pathogen. Chlorothalonil (1,250 g ai/ha dose), and four other fungicides inhibited infectivity and prevented mortality to aphids, following their exposure to the fungus. Mueller et al. (2005) observed that chlorothalonil, and 12 other fungicides eliminated the germination of 6 rust fungi (*Puccinia hemerocallidis, P. iridis, P. menthae, P. oxalis, P. pelargonii-zonalis, and Pucciniastrum vaccinii*) within 24 h, when they were exposed during and after fungicide application; chlorothalonil completely inhibited spore germination within 8 h.

6 Summary

Chlorothalonil is a broad spectrum, non systemic, organochlorine pesticide that was first registered in 1966 for turfgrasses, and later for several food crops. Chlorthalonil has both a low Henry's law constant and vapor pressure, and hence, volatilization losses are limited. Although, chlorothalonil's water solubility is low, studies have shown it to be highly toxic to aquatic species. Mammalian toxicity (to rats and mice) is moderate, and produces adverse effects such as, tumors, eye irritation and weakness. Although, there is no indication that chlorothalonil is a human carcinogen, there is sufficient evidence from animal studies to classify it as a probable carcinogen.

Chlorothalonil has a relatively low water solubility and is stable to hydrolysis. However, hydrolysis under basic conditions may occur and is considered to be a minor dissipation pathway. As a result of its high soil adsorption coefficient this fungicide strongly sorbs to soil and sediment. Therefore, groundwater contamination is minimal. Degradation via direct aqueous or foliar photolysis represents a major dissipation pathway for this molecule, and the photolysis rate is enhanced by natural photosensitizers such as dissolved organic matter or nitrate. In addition to photolysis, transformation by aerobic and anaerobic microbes is also a major degradation pathway. Under anaerobic conditions, hydrolytic dechlorination produces the stable metabolite 4-hydroxy-2,5,6-trichloroisophthalonitrile. Chlorothalonil is more efficiently degraded under neutral pH conditions and in soil containing a low carbon content.

Acknowledgments Support was provided by the Environmental Monitoring Branch of the California Department of Pesticide Regulation (CDPR), California Environmental Protection Agency, under contract No. 10-C0102. The statements and conclusions are those of the authors and not necessarily those of CDPR. The mention of commercial products, their source, or their use in connection with materials reported herein is not to be construed as actual or implied endorsement of such products. Special thanks to Kean Goh for his assistance.

References

Armbrust K (2001) Chlorothalonil and chlorpyrifos degradation products in golf course leachate. Pest Manag Sci 57:797–802

Bedos C, Rousseau-Djabri MF, Loubet B, Durand B, Flura D, Briand O, Barriuso E (2010) Fungicide volatilization measurements: inverse modeling, role of vapor pressure, and state of foliar residue. Environ Sci Technol 44:2522–2528

Bringolf RB, Cope WG, Eads CB, Lazaro PR (2007) Acute and chronic toxicity of technical-grade pesticides to glochidia and juveniles of freshwater mussels (unionidae). Environ Toxicol Chem 26(10):2086–2093

CDPR, California department of Pesticide Regulation (2005) Chlorothalonil: risk characterization document for dietary exposure. http://www.cdpr.ca.gov/docs/risk/rcd/chlorothalonil.pdf

Caux PY, Kent RA, Fan GT, Stephenson GL (1996) Environmental fate and effects of chlorothalonil: a Canadian perspective. Crit Rev Environ Sci Tech 26(1):45–93

Chen SK, Edwards CA, Subler S (2001) Effects of the fungicides benomyl, captan and chlorothalonil on soil microbial activity and nitrogen dynamics in laboratory incubations. Soil Biol Biochem 33:1971–1980

Carlo-Rojas Z, Bello-Mendoza R, Figueroa MS, Sokolov MY (2004) Chlorothalonil degradation under anaerobic conditions in an agricultural tropical soil. Water Air Soil Pollut 151:397–409

Davies PE, White RWG (1985) The toxicology and metabolism of chlorothalonil in fish. I. Lethal levels for *Salmo gairdneri*, *Galaxias maculatus*, *G. truttaceus* and *G. auratus* and the fate of 14C-TCIN in *S. gairdneri*. Aquat Toxicol 7:93–105

DeLorenzo ME, Wallace SC, Danese LE, Baird TD (2009) Temperature and salinity effects on the toxicity of common pesticides to the grass shrimp, *Palaemonetes pugio*. J Environ Sci Health B 44:455–460

Ernst W, Doe K, Jonah P, Young J, Julien G, Hennigar P (1991) The toxicity of chlorothalonil to aquatic fauna and the impact of its operational use on a pond ecosystem. Arch Environ Contam Toxicol 21:1–9

Farag AT, Abdel-Zaher Karkour T, El Okazy A (2006) Embryotoxicity of oral administered chlorothalonil in mice. Birth Defect Res B 77:104–109

Fungicide Resistance Action Committee (2013) FRAC code list 2013: fungicides sorted by mode of action

Fushiwaki Y, Urano K (2001) Adsorption of pesticides and their biodegradation products on clay minerals and soils. J Health Sci 47(4):429–432

Gallagher EP, Canada AT, Di Giulio RT (1992) The protective role of glutathione in chlorothalonil-induced toxicity to channel catfish. Aquat Toxicol 23:155–168

Gamble DS, Bruccoleri AG, Lindsay E, Langford AH (2000) Chlorothalonil in a quartz sand soil: speciation and kinetics. Environ Sci Technol 34:120–124

Habte M, Aziz T, Yuen JE (1992) Residual toxicity of soil-applied chlorothalonil on mycorrhizal symbiosis in *Leucaena leucocephala*. Plant Soil 140:263–268

Haith DA, Rossi FS (2003) Risk assessment of pesticide runoff from turf. J Environ Qual 32:447–455

Kwon JW, Armbrust KL (2006) Degradation of chlorothalonil in irradiated water/sediment systems. J Agric Food Chem 54:3651–3657

Latteur G, Jansen JP (2002) Effects of 20 fungicides on the infectivity of conidia of the aphid entomopathogenic fungus *Erynia neoaphidis*. BioControl 47:435–444

Leistra M, Van Den Berg F (2007) Volatilization of parathion and chlorothalonil from a potato crop simulated by the PEARL model. Environ Sci Technol 41:2243–2248

Liang B, Li R, Jiang D, Sun J, Qiu J, Zhao Y, Li S, Jiang J (2010) Hydrolytic dechlorination of chlorothalonil by *Ochrobactrum* sp. CTN-11 isolated from a chlorothalonil-contaminated soil. Curr Microbiol 61:226–233

Monadjemi S, El Roz M, Richard C, Ter Halle A (2011) Photoreduction of chlorothalonil fungicide on plant leaf models. Environ Sci Technol 45:9582–9589

Mori T, Fujie K, Kuwatsuka S, Katayama A (1996) Accelerated microbial degradation of chlorothalonil in soils amended with farmyard manure. Soil Sci Plant Nutr 42(2):315–322

Motonaga K, Takagi K, Matumoto S (1996) Biodegradation of chlorothalonil in soil after suppression of degradation. Biol Fertil Soils 23:340–345

Mozzachio AM, Rusiecki JA, Hoppin JA, Mahajan R, Patel R, Beane-Freeman L, Alavanja MCR (2008) Chlorothalonil exposure and cancer incidence among pesticide applicator participants in the agricultural health study. Environ Res 108:400–403

Mueller DS, Jeffers SN, Buck JW (2005) Toxicity of fungicides to urediniospores of six rust fungi that occur on ornamental crops. Plant Dis 89:255–261

Park J-W, Lee S-E, Rhee I-K, Kim J-E (2002) Transformation of the fungicide chlorothalonil by fenton reagent. J Agric Food Chem 50:7570–7575

Patakioutas G, Albanis TA (2002) Adsorption-desorption studies of alachlor, metolachlor, EPTC, chlorothalonil and pirimiphos-methyl in contrasting soils. Pest Manag Sci 58:352–362

Penuela GA, Barcelo D (1998) Photodegradation and stability of chlorothalonil in water studied by soild-phase disk extraction followed by gas chromatographic techniques. J Chromatogr A 823:81–90

Potter TL, Wauchope RD, Culbreath AK (2001) Accumulation and decay of chlorothalonil and selected metabolites in surface soil following foliar application to peanuts. Environ Sci Technol 35:2634–2639

Putnam RA, Nelson JO, Clark JM (2003) The persistence and degradation of chlorothaonil and chlorpyrifos in a cranberry bog. J Agric Food Chem 51:170–176

Sakkas VA, Lambropoulou DA, Albanis TA (2002) Study of chlorothalonil photodegradation in natural waters and in the presence of humic substances. Chemosphere 48(9):939–945

Sapozhnikova Y, Wirth E, Schiff K, Brown J, Fulton M (2007) Antifouling pesticides in the coastal waters of Southern California. Mar Pollut Bull 54:1972–1978

Sato K, Tanaka H (1987) Degradation and metabolism of a fungicide, 2,4,5,6-tetrechloroisophthalonitrile (TPN) in soil. Biol Fertil Soils 3:205–209

Sherrard RM, Murray-Gulde CL, Rodgers JH, Shah YT (2003) Comparative toxicity of chlorothalonil: *Ceriodaphnia dubia* and *Pimephales promelas*. Ecotoxicol Environ Saf 56:327–333

Szalkowski MB, Stallard DE (1977) Effect of pH on the hydrolysis of chlorothalonil. J Agric Food Chem 25(1):208–210

Tillman RW, Siegel MR, Long JW (1973) Mechanism of action and fate of the fungicide chlorothalonil (2,4,5,6-tetrachloroisophthalonitrile) in biological systems. I. Reactions with cells and subcellular components of *Saccharomyces pastorianus*. Pest Biochem Physiol 3:160–167

Tomlin CDS (2000) The pesticide manual, 12th edn. The British Crop Protection Council, Surrey, UK, pp 620–621

Ukai T, Itou T, Katayama A (2003) Degradation of chlorothalonil in soils treated repeatedly with chlorothalonil. J Pest Sci 28:208–211

United States Environmental Protection Agency. Office of Pesticide Programs. Special Review and Reregistration Division., Reregistration eligibility decision: chlorothalonil (1999) US Environmental Protection Agency Office of Pesticide Programs Special Review and Reregistration Division: Washington, D.C.

United States Environmental Protection Agency (2007) Office of Pesticide Programs. Potential risks of labeled chlorothalonil uses to the federally listed California red legged frog. 2007, US Environmental Protection Agency Office of Pesticide Programs Environmental Fate and Effects Division: Washington, D.C.

USGS National Water Quality Assessment Data Warehouse http://web1.er.usgs.gov/NAWQAMapTheme/index.jsp

van der Pas LJT, Matser AM, Boesten JJTI, Leistra M (1999) Behaviour of metamitron and hydroxychlorothalonil in low-humic sandy soils. Pest Sci 55:923–934

Vincent PG, Sisler HD (1968) Mechanism of antifungal action of 2,4,5,6-tetrachloroisophthalonitrile. Physiol Plant 21:1249–1264

Wallace DF, Hand LH, Oliver RG (2010) The role of indirect photolysis in limiting the persistence of crop protection products in surface waters. Environ Toxicol Chem 29(3):575–581

Waltz C, Armbrust K, Landry G (2002) Chlorpyrifos and chlorothalonil in golf course leachate. http://www2.gcsaa.org/gcm/2002/sept02/pdfs/09chlorpyrifos.pdf

Wan MT, Rahe JE, Watts RG (1998) A new technique for determining the sublethal toxicity of pesticides to the vesicular-arbuscular mycorrhizal fungus *Glomus Intraradices*. Environ Toxicol Chem 17(7):1421–1428

Wang H, Xu S, Hu J, Wang X (2009) Effect of potassium dihydrogen phosphate and bovine manure compost on the degradation of chlorothalonil in soil. Soil Sediment Contam 18:195–204

Wang H, Wang C, Chen F, Wang X (2011) Anaerobic degradation of chlorothalonil in four paddy soils. Ecotoxicol Environ Saf 74:1000–1005

World Health Organization (1996) International Programme on Chemical Safety. Chlorothalonil. Environmental Health Criteria 183. Geneva, Switzerland. http://www.inchem.org/documents/ehc/ehc/ehc183.htm#SubSectionNumber:9.1.3

Wu L, Liu G, Yates MV, Green RL, Pacheco P, Gan J, Yates SR (2002) Environmental fate of metalaxyl and chlorothalonil applied to a bentgrass putting green under southern California climatic conditions. Pest Manag Sci 58:335–342

Zhang XH, Zhu YG, Lin AJ, Chen BD, Smith SE, Smith FA (2006) Arbuscular mycorrhizal fungi can alleviate the adverse effects of chlorothalonil on Oryza sativa L. Chemosphere 64:1627–1632

Zhang Y, Lu J, Wu L, Chang A, Frankenberger WT (2007) Simultaneous removal of chlorothalonil and nitrate by Bacillus cereus strain NS1. Sci Total Environ 382:383–387

The Distribution, Fate, and Effects of Propylene Glycol Substances in the Environment

Robert West, Marcy Banton, Jing Hu, and Joanna Klapacz

Contents

1 Introduction 108
2 Physico-Chemical Properties 109
    2.1 Density (Specific Gravity) 110
    2.2 Melting/Freezing Point 111
    2.3 Boiling Point 111
    2.4 Vapor Pressure 111
    2.5 Water Solubility 112
    2.6 Henry's Law Constant 112
    2.7 Octanol-Water Partition Coefficient (Log P_{ow}) 113
    2.8 Organic Carbon-Normalized Adsorption Coefficient (Log K_{oc}) 113
3 Environmental Distribution 114
    3.1 Relevant Environmental Compartment(s) 114
    3.2 Environmental Monitoring Data 117
4 Environmental Fate Processes 119
    4.1 Atmospheric Fate/Transport 119
    4.2 Biodegradation 120
    4.3 Hydrolysis 124
    4.4 Bioaccumulation 125
5 Ecotoxicity 126
    5.1 Monopropylene Glycol (MPG) 126
    5.2 Dipropylene Glycol (DPG) 126
    5.3 Tripropylene Glycol (TPG) 130
    5.4 Tetrapropylene Glycol (TePG) and Higher Oligomers 130
6 Potential for Endocrine Disruption 132
7 Summary 133
References 134

R. West (✉) • J. Hu • J. Klapacz
Toxicology and Environmental Research & Consulting (TERC), The Dow Chemical Company, 1803 Building, Midland, MI, USA
e-mail: rjwest@dow.com

M. Banton
Lyondell Chemical Company, 1221 McKinney Street, Suite 700, Houston, TX, USA

1 Introduction

The family of synthetic organic substances known as "propylene glycols" consists of the 1,2-propanediol substance (monopropylene glycol, MPG) and its dimer (dipropylene glycol, DPG), trimer (tripropylene glycol, TPG) and tetramer (tetrapropylene glycol, TePG) forms. The formal identities of these substances are summarized in Table 1. Collectively, these substances are produced on a scale of approximately three million metric tons per year, and are among the most important group of synthetic organic chemicals in commerce today (Chinn and Kumamoto 2011). Produced and used globally, the propylene glycol (PG) substances have functional properties that enable their application in the manufacture of polyester resins and their formulation into functional fluids (*e.g.*, anti-freeze, aircraft anti-icing and de-icing fluids), cosmetics, pharmaceuticals, personal care products, pesticides, liquid detergents, paints and coatings, and foods used for human and animal consumption. The PG substances also have more minor uses as a humectant for tobacco, plasticizers, and solvents used in fragrance, agricultural and ink formulations. Considering the sheer volume consumed in these broad and dispersive applications, a variety of scenarios can be envisioned for their emission to the environment. Thus, there is a need to understand the potential hazards of and exposures associated with the manufacture, transport, use and disposal of products containing or manufactured from the PG substances.

The purpose of this review is to summarize and communicate the best-available information to enable assessments of hazard, exposure and risk that are associated with the PG substances over their life cycle stages, which involve direct or diffusive environmental emission. Although various technical mixtures of these PG substances are of commercial and regulatory interest, the distribution, fate of and exposures to these mixtures in the environment are determined by the properties of each individual PG substance, rather than by the composition or properties of the collective mixture.

Table 1 Identity of the propylene glycol substances and associated components

Common name (abbreviated name)	Chemical abstracts services registry		Purity as tested for physico-chemical properties
	Name	Number	
Propylene glycol (MPG)	1,2-Propanediol	57-55-6	99.90%
Dipropylene glycol (DPG)	Propanol, oxybis	25265-71-8	99.77%
	2-Propanol, 1,1,-oxybis-	110-95-8	
	1-Propanol, 2,2'-oxybis-	108-61-2	
	1-Propanol,2-(2-hydroxypropoxy)	106-62-7	
Tripropylene glycol (TPG)	Propanol, [(1-methyl-1,2-ethanediyl)bis(oxy)]bis-	24800-44-0	≥99.40%
Tetrapropylene glycol (TePG)	1-Propanol, 2-[2-[2-(2-hydroxypropoxy)propoxy]propoxy]-	24800-25-7	≥99.7%
	Propanol, [oxybis[(methyl-2,1-ethanediyl)oxy]]bis-	25657-08-3	

A series of polymeric PG substances (*i.e.*, poly(propylene glycol) or PPG) are commercially prepared over a wide range of average molecular weights, and these have equally varied properties that are distinct from those of the MPG, DPG, TPG and TePG homologues. Where available, fate and ecological hazard information is presented here for the low molecular weight polymers having number-average molecular weight (M_n) ≤500 g/mol, the components of which may include the MPG–TePG homologues. Therefore, the foci of this review are the physical/chemical, fate and ecotoxicological properties that influence the distribution and exposure of these four individual substances in the environment. A separate review has been recently completed, in which the physical/chemical and toxicological hazards that are associated with potential human exposures to these substances are summarized (Fowles et al. 2013). In that review, the identities, structures, and compositions associated with the commercial PG substances are also detailed, and are therefore not revisited here.

2 Physico-Chemical Properties

The following physico-chemical properties influence the emission, transport, and fate of PG substances in the environment: melting (freezing) point, density, boiling point, vapor pressure, water solubility, octanol-water partition coefficient (log P_{ow}), and organic carbon-normalized partition coefficient (log K_{oc}). Considering that the PG substances have been in commerce for many decades, these and other properties have been measured numerous times for various purposes, and are reported in various secondary reference sources, wherein details of the measurement techniques are often lacking (Brown et al. 1980; Puck and Tamplin 1952; Sullivan 1993; Verschueren 2001; Weast and Astle 1985). Various processes were used to commercially manufacture and isolate these substances, and different processes yield the potential for introduction of different impurities and co-products. Neither the effect of such different impurities and co-products on physical/chemical property measurements, nor the reliability and relevance of these previously-reported properties can be fully ascertained. Therefore, in partial fulfillment of the substance registration requirements under the European Union REACh legislation (Regulation (EC) No 1907/2006 of The European Parliament and of The Council of 18 December 2006), significant effort and expense were undertaken to measure selected physical/chemical properties for highly-purified and characterized samples of the MPG, DPG, and TPG substances. These measurements followed current and globally-accepted standardized test procedures as put forth by OECD (OECD 2012) and the European Community (EC 2008). The measurements relied on laboratory procedures and test substance characterizations performed under the OECD principles of Good Laboratory Practice (OECD 2003), so that the purity of the tested substances and reliability of associated results can be determined. The results of these characterizations are summarized in Table 2, and are expected to represent the most accurate, reliable, and traceable physico-chemical property data available for assessing the environmental distribution and fate of PG substances.

Table 2 Summary of selected physical/chemical properties of the propylene glycol substances

Property	MPG	DPG	TPG	TePG[a]
Molar mass (g/mol)	76.10	134.18	192.26	250.34
Density (g/cm³ @ 20 °C)	1.03	1.02	1.02	1.02
Freezing point (°C)	<−20	<−20	<−20	−7.3
	−60[a]	−40[a]	−45[a]	
Boiling point (°C)	184	227	270	312
Vapor pressure (Pa @ 101.3 kPa and 25 °C)	20	1.3	0.26	0.0036
Water solubility	---------------Miscible---------------			
Henry's law constant (Pa m³/mol)	1.3×10^{-3}	1.8×10^{-4}	5.0×10^{-5}	4.7×10^{-7}
Log P_{ow}	−1.07	−0.46	−0.38	−0.35[b]
Log K_{oc}[b]	−0.49	−0.24	−0.29	−0.51

[a]Critically-reviewed and recommended values from the AIChE DIPPR Database (AIChE 2012)
[b]Estimated value based on KOCWIN and KOWWIN software (USEPA 2012)

It should be noted that the TePG substance, in its purified commercial form, did not meet the import/production tonnage trigger (>1,000 t/year) for REACh registration in 2010. Other technical mixtures containing TePG often meet the OECD definition of a polymer (OECD 1991), and as such, are not subject to the registration requirements of REACh. Therefore, there was no regulatory need to re-assess the physico-chemical properties of TePG. However, the available measurements of density, freezing point, boiling point, and vapor pressure for this substance have been critically-reviewed under the American Institute of Chemical Engineers (AIChE) Design Institute for Physical Properties Research (DIPPR) program (AIChE 2012), and from this source the recommended values are reported in Table 2. Similarly, the critically-evaluated and accepted property values reported in DIPPR for the MPG, DPG, and TPG substances are in excellent agreement with the most recently-measured values that were determined for REACh registration (Table 2).

2.1 Density (Specific Gravity)

The density of a substance is an important property that can influence how a released substance migrates within and among air, water, and soil. For example, a spillage of bulk liquid to surface water can result in that substance floating, sinking, or remaining suspended in the receiving water body. Pure distilled water has a density of 0.998 g/cm³ at 20 °C (Landolt-Bornstein 1980), whereas water from the open ocean has typical density of approximately 1.025 g/cm³ at the same temperature and at 3.5% salinity (Cox et al. 1970). The PG substances exhibit a very narrow range of densities, from 1.02 to 1.03 g/cm³ at 20 °C, and if spilled to surface waters would tend to float or slowly sink until readily and completely dissolved. For this reason, there are no practical measures for recovering PG substances from surface waters following their bulk spillage.

2.2 Melting/Freezing Point

The melting/freezing point indicates whether a substance occurs as a solid or liquid at standard atmospheric pressure (101.325 kPa), and at a given temperature associated with processing, use, or emission. Because each of the PG substances occurs as a viscous liquid at ambient temperature (25 °C), this change in physical state between liquid and solid, which may occur at lower temperatures, is expressed as the freezing point. Measurement of the freezing points for MPG, DPG, and TPG were attempted by using the differential scanning calorimetry (DSC) procedure described in the EC Method A.1 (EC 2008). In this procedure, the heat flow into or out of the sample of test substance is measured as the sample is slowly cooled to a minimum temperature of −20 °C. For each of the MPG, DPG, and TPG substances, an exothermic change of state was not observed for temperatures as low as −20 °C. Therefore, the freezing point for these substances is reported as <−20 °C in Table 2. Critically-evaluated and accepted measurements of freezing point, as reported in the DIPPR database, indicate that this transition to glassy solid (*i.e.*, glass transition temperature) occurs at temperatures as low as −60 °C (MPG) and as high as −7.3 °C (TePG). Therefore, the PG substances, in their pure forms, will occur as flowable viscous liquids at virtually any temperature associated with manufacture, transport, storage or use.

2.3 Boiling Point

The boiling point indicates the temperature at which the pure PG substances will change from liquid to gas (vapor) state at standard atmospheric pressure (101.325 kPa). As with the freezing point, this property is used in multimedia models to determine the physical state of a substance at a given ambient environmental temperature. For the MPG, DPG, and TPG substances, the reported boiling points were determined by using the differential scanning calorimetry technique of the EC Method A.2 (EC 2008), which indicates the onset of the endothermic phase transition as temperature of a sample is increased incrementally. The PG substances, having normal boiling points ranging from 184 to 312 °C, will remain in the liquid state at any temperature associated with their use or emission to the environment. It is also important to note that, during boiling point measurements, the thermal decomposition of the PG substances is not observed, which is indicative of their high degree of thermal stability.

2.4 Vapor Pressure

The vapor pressure of a substance indicates the fraction of a substance that exists in the vapor phase at a given temperature, and is typically measured and reported for temperatures of 20 or 25 °C. When a substance occurs in a neat liquid form, the vapor pressure indicates the propensity of that substance to volatilize into the atmosphere. The PG substances exhibit a wide range of measured vapor pressures, from 0.0036 to 20 Pa at 25 °C (Table 2); however, this range of vapor pressure can be characterized as indicating low volatility of the substances in their pure forms. Although butyl acetate is assigned an evaporation rate of 1.0, the most volatile PG substance (*i.e.*, MPG) has a relative evaporation rate of 0.016, and that of the least-volatile (TPG) is 0.0002 (The Dow Chemical Company 2003). These substances are therefore considered to have low evaporation rates, considering that water has a relative evaporation rate of 0.3.

2.5 Water Solubility

The water solubility of a substance determines, in part, the limit to which mass transfer (advection) of that substance can occur when it is dissolved in surface- or ground-water. It also indicates the extent to which wet deposition of the substance vapor or aerosols can occur from the atmosphere. The water-soluble fraction of a substance is most susceptible to degradation reactions, such as biodegradation, hydrolysis, and photolysis. The PG substances are each reported to be miscible with water in all proportions; however, the rate with which the substances will dissolve in water apparently decreases as molecular weight and associated viscosity increases. Although many higher molecular weight glycol substances exhibit inverse solubility (*i.e.*, decreased solubility with increased temperature), aqueous solutions of the PG substances are expected to remain fully dissolved at temperatures up to and including their boiling points. Thus, the distribution, transport, degradation, and toxicity of the PG substances in the environment will not be limited by their water solubility.

2.6 Henry's Law Constant

When a substance is dissolved in water, both the vapor pressure and water solubility of the substance will determine the degree to which it will volatilize to the atmosphere. The quotient of vapor pressure (Pascal) and water solubility (mol/m^3) provides an estimate this volatility from water, as described by Henry's Law Constant (Pa m^3/mol). Thus, Henry's Law Constant (HLC) indicates the degree to which partitioning of a substance occurs between dissolved and vapor phases for an aqueous solution at a given temperature. The HLC can be directly measured, by determining concentrations of the substance in dissolved aqueous and vapor phases of equilibrated solutions in closed vessels. More often, the HLC is estimated from the quotient of measured or estimated vapor pressure (Pa) and water solubility (mol/m^3). By this estimation method, HLC values for the PG substances range from 1.3×10^{-3} Pa m^3/mol (MPG) to 4.7×10^{-7} Pa m^3/mol (TePG) at 25 °C. These HLC values indicate that the PG substances are poorly-, to essentially non-volatile, from water. Accordingly, any emissions to surface water or soil will not tend to be volatilized to the atmosphere. Rather, any atmospheric emissions (vapor or aerosol) will tend to be readily deposited to water or soil by wet deposition.

2.7 Octanol-Water Partition Coefficient (Log P_{ow})

The octanol-water partition coefficient, often expressed in its base-10 logarithm form (*i.e.*, log P_{ow}) is among the most important properties describing the fate and distribution of a substance in the environment. When octanol is employed as a surrogate for fatty tissues (*i.e.*, lipids), the log P_{ow} is highly-correlated with the bioconcentration of substances in aquatic organisms (Dimitrov et al. 2005). When octanol is employed as a surrogate for natural organic matter in soil, sediment, or wastewater treatment bio-solids, the log P_{ow} is highly-correlated with the organic carbon-normalized adsorption coefficient, or log K_{oc} (Lyman 1990; Sabljic et al. 2005). The log P_{ow} values for MPG, DPG, and TPG have been determined for highly-purified forms of these substances, according to EC Method A.8 (EC 2008). In this method, the concentration of the substance (including all structural and stereo-isomers) is determined in both the water and 1-octanol phases of equilibrated and mutually-saturated octanol/water mixtures prepared at three different octanol:water (vol:vol) ratios. The log P_{ow} values measured as such for the MPG, DPG, and TPG substances range from −1.07 to −0.38. Because a measured value of log P_{ow} is not available for the TePG substance, the estimated value of −0.35 is reported, which originates from the widely accepted and validated structure-fragment calculation software KOWWIN v1.68 (USEPA 2012).

From this series of measured and calculated values of log P_{ow} for the PG substances, it is clear that the addition of each oxypropylene repeat unit to propylene glycol makes a net positive (*i.e.*, hydrophobic) contribution to log P_{ow} of the higher PG homologues. Although the measured log P_{ow} values for MPG, DPG, and TPG do not indicate a uniform contribution to log P_{ow} from the oxypropylene repeat unit, the structure-fragment calculation method indicates a uniform contribution of approximately +0.14 log P_{ow} units. The calculated log P_{ow} values for MPG, DPG, TPG, and TePG are −0.78, −0.64, −0.50, and −0.35, respectively. Regulatory criteria are based on log P_{ow}, and typically are derived from a screening assessment of bioaccumulation potential, where a log P_{ow} value ≥3 is the lowest threshold applied (by the International Maritime Organization) to indicate a potential for bioaccumulation (Moermond et al. 2011). In all cases, whether log P_{ow} is measured or calculated, very low potentials for bioaccumulation and adsorption to soil, sediment, or wastewater bio-solids are indicated for the PG substances.

2.8 Organic Carbon-Normalized Adsorption Coefficient (Log K_{oc})

As described above, the degree to which organic matter in soil, sediment, and wastewater bio-solids adsorbs non-ionic organic substances is indicated by the organic carbon-normalized adsorption coefficient (*i.e.*, log K_{oc}). For the PG substances, no reported or traceable measured values of log K_{oc} exist. However, several techniques exist that utilize correlations with log P_{ow} or molecular connectivity indices to estimate log K_{oc} values for these substances. The KOCWIN software v2.00 (USEPA 2012) provides estimates that are based on both of these techniques, and is among the most widely-recognized and accepted tools for estimating log K_{oc} of non-ionic organic substances. When a reliable measured value for log P_{ow} exists, the log K_{oc} estimate is most accurately made by correlation with log P_{ow}. When the log P_{ow} value is unknown, or the substance possesses ionizable functional groups, the molecular connectivity index (MCI) may serve to provide a more accurate log K_{oc} estimate. The estimated values of log K_{oc} for the MPG, DPG, TPG, and TePG substances, resulting from the log P_{ow} correlation method of KOCWIN, are −0.49, −0.24, −0.29, and −0.51, respectively (Table 2). Because each estimate is corrected for various structural or molecular features, these estimates do not show a uniform and incremental increase in log K_{oc} with each additional oxypropylene repeat unit. Substances having a log K_{oc} value <3 are considered to be poorly adsorbed to sediment and soil (SETAC 1993), such that assessments of their potential toxicity to sediment-dwelling organisms are not typically undertaken. To summarize the foregoing, the PG substances have very low potential for adsorption to soil, sediment, and wastewater bio-solids, and their advection into and through groundwater will not be appreciably attenuated by adsorption.

3 Environmental Distribution

The physico-chemical properties of a substance influence its distribution and fate in the environment, as well as the route by which the substance is emitted to the environment. For example, tetrachloroethylene has high vapor pressure (2,415 Pa) and only moderate water solubility (150 mg/L) at 25 °C (ECHA 2013a), and might be expected to occur primarily in the atmospheric compartment of the environment. However, this substance has widespread occurrence as a groundwater contaminant, because past use and disposal practices resulted in its direct emission to surface soils (Moran and Delzer 2006). Thus, to understand or predict where a substance might reside in the environment, both the physico-chemical properties and modes of emission must be well-understood and considered.

3.1 Relevant Environmental Compartment(s)

The Level III fugacity-based multimedia fate and transport model, developed by Don Mackay and colleagues (Mackay 2001), provides a convenient and meaningful approach to identifying the relevant environmental compartments associated with environmental emissions of a substance. This model determines the steady-state concentrations of a substance in the modeled environmental compartments, under various simulated modes and magnitudes of continuous emission. The inputs to this model include physico-chemical properties as discussed above and as summarized in Table 2, known or hypothetical route(s) and magnitude(s) of emission, and estimated degradation half-lives for the substance in the atmospheric, water, soil, and sediment

Table 3 Estimated environmental degradation half-lives used in Level III distribution modeling for the propylene glycol substances

Substance	Second-order reaction rate constant (cm^3/molecule*s) with OH radical @ 25 °C[a]	Estimated half-life (h)			
		Atmosphere[b]	Soil	Water	Sediment
MPG	1.3×10^{-11}	10	720	360	720
	1.2×10^{-11c}	10.7[c]			
DPG	$(3.1–3.4) \times 10^{-11}$	3.7–4.1	720	360	720
TPG	$(5.6–5.9) \times 10^{-11}$	2.2–2.3	720	360	720
TePG	$(7.5–8.1) \times 10^{-11}$	1.6–1.7	1,440	720	1,440

[a]Rate constant estimated from structure fragment correlation method of AOPWIN v1.92a (USEPA 2012)
[b]Based on an assumed average hydroxyl radical concentration of 1.5×10^6 molecules/cm^3
[c]Experimentally-measured value of Atkinson (1986)

compartments (Table 3). The estimated half-lives for the substances in the atmosphere were derived from estimated second-order reaction rate constants, as described in Sect. 4.1. The estimated half-lives in soil, water, and sediment are derived from the demonstrated ready biodegradability (MPG, DPG, TPG) and inherent ultimate biodegradability (TePG) of the substances, their estimated soil adsorption coefficients (log Koc, Table 2), and the corresponding default half-lives recommended under the U.S. EPA High Production Volume Chemicals program (Larson et al. 2000). The Level III model (v2.80.1; CEMC 2004) was used to identify the relevant environmental compartment(s) that are associated with various routes of emission for the PG substances. For each of these four substances, four different emission scenarios were evaluated, with 1,000 kg/h emissions (both individually and simultaneously) to the air, water, soil compartments of the standard "EQC" model environment (Mackay 2001). The resulting predicted distributions of the emitted PG substance in each compartment, and their associated residence times in the total environment, are summarized in Table 4a–d. For each of the four emission scenarios, the predicted percentage of total steady-state mass of PG substance occurring in the air, water, soil, and sediment compartments is given. Moreover, the residence time (day) over which a given molecule of the PG substance occurs in the environment is predicted for conditions under which that molecule is removed from the environment by advection only, and by the combined effects of advection and degradation processes.

The results of the Level III modeling illustrate several key and expected behaviors of the PG substances in the environment. As discussed above, because these substances have low vapor pressures and very high water solubility, they are not expected to reside in the atmosphere, regardless of the route by which they reach the environment. Even if emitted directly to the atmosphere, each PG substance is predicted to be completely deposited to surface water and soil, in the same approximate proportion as exists for the surface areas of water and soil compartments in the simulated environment. Thus, wet deposition of the PG substances would appear to be an important fate process affecting any atmospheric emissions. The simulated emission of these substances directly to surface waters is predicted to result in their

Table 4 Summary of Level III model-predicted environmental distributions and residence times associated with simulated emissions of the propylene glycol substances

Emission scenario	Predicted distribution (%) in:				Residence time (days)	
	Atmosphere	Water	Soil	Sediment	Advection	Total
(a) Monopropylene glycol (MPG)						
1,000 kg/h Atmosphere	0.9	25.9	73.2	0.0	119	19.1
1,000 kg/h Water	0.0	99.9	0.0	0.1	41.7	14.3
1,000 kg/h Soil	0.0	22.2	77.8	0.0	187	29.6
1,000 kg/h atm, water, and soil	0.3	40.9	58.8	0.1	95.3	21
(b) Dipropylene glycol (DPG)						
1,000 kg/h Atmosphere	0.1	25.9	74.0	0.0	155	25.1
1,000 kg/h Water	0.0	99.9	0.0	0.1	41.7	14.3
1,000 kg/h Soil	0.0	22.0	78.0	0.0	190	29.9
1,000 kg/h atm, water, and soil	0.0	39.4	60.5	0.1	105	23.1
(c) Tripropylene glycol (TPG)						
1,000 kg/h Atmosphere	0.0	25.8	74.1	0.0	159	26.7
1,000 kg/h Water	0.0	99.9	0.0	0.1	41.7	14.3
1,000 kg/h Soil	0.0	21.9	78.0	0.0	190	29.9
1,000 kg/h atm, water, and soil	0.0	39.1	60.8	0.1	106.0	23.6
(d) Tetrapropylene glycol (TePG)						
1,000 kg/h Atmosphere	0.0	33.0	67.0	0.1	126	42.9
1,000 kg/h Water	0.0	99.8	0.0	0.2	41.7	21.3
1,000 kg/h Soil	0.0	29.5	70.5	0.0	141	45.4
1,000 kg/h atm, water, and soil	0.0	44.5	55.4	0.1	93.7	36.5

retention in the surface water compartment, with virtually no evaporation to the atmosphere or deposition to sediments. When emitted to soil, the PG substances will become associated almost exclusively with soil pore water, and will have, approximately, a 20–30% runoff to surface waters.

The PG substances are rapidly degraded in air, water, soil, and sediment (as discussed below); hence, their residence times in the environment are expected to be governed by their degradation rate more than by advection. The degradation half-life times (h) input to the model for indirect photolysis of each PG substance in the atmosphere are summarized in Table 3. Degradation half-lives in surface water, soil, and sediment compartments for the PG, DPG, and TPG substances were 360, 720, and

720 h, respectively, as recommended for derivation of biodegradation half-life times from results of readily biodegradability tests (Larson et al. 2000). Similarly, for inherently biodegradable substances, the input half-life times in these media for the TePG substance were derived as 720, 1,440, and 1,440 h for water, soil, and sediment, respectively. As shown in Table 4a–d, the total residence times for all substances and emission scenarios range from 14 to 45 days. When the reactivity of the model is turned "off", and only advection is allowed to govern the fate and transport in and through the environment, predicted residence times range from 41.7 to 190 days. Using the Level III model in this way illustrates the importance of reactivity (degradation) of the PG substances in governing their environmental fate and transport.

From the Level III modeling, it can be concluded that the surface water environment is of primary interest when addressing the fate and effects of the PG substances, regardless of the mode by which the substances might be emitted. The soil environment is expected to be of interest only when these substances are emitted directly to soil, or are deposited there from continuous atmospheric emissions during their manufacture, transport, or use. Thus, the focus of environmental hazard assessments for these substances should be aquatic and terrestrial organisms at all trophic levels.

3.2 Environmental Monitoring Data

Various voluntary and regulatory-mandated programs have been implemented through which the presence of chemical substances, especially those of high hazard and/or production volume, are monitored in samples collected from air, water, soil, sediment, and biota. The results of these environmental monitoring programs can provide a useful check on effectiveness of waste treatment processes, emission controls, and disposal practices that are associated with manufacture, use, and disposal of these substances.

Searches of the published literature, government databases, and internet sources have revealed very little information on detection of the PG substances in the environment, or to monitoring programs that have included the PG substances as target analytes. The OECD SIDS Initial Assessment Reports (SIAR) compiled for PG (OECD 2001a) and TPG (OECD 1994), which have sections that address environmental monitoring information, include no information on detection of these substances in air, water, soil, or sediment. The SIAR report for DPG (OECD 2001b) included reports of detections in drinking water (0.2–0.4 ng/L; Lin et al. 1981), pulp/paper mill wastewater effluent (11 µg/L; Turoski et al. 1983), and ground water (Dunlap and Shew 1976). Because the OECD SIDS program has now concluded without sponsorship of the TePG substance, a similar SIAR report is not available for this substance.

Various local surface- and ground-water monitoring programs have been established at airport facilities that use aircraft de-icing and/or anti-icing formulations, which can contain up to 90% MPG (The Dow Chemical Company 2013). For example, Sills and Blakeslee (1992) reviewed available information on the environmental impact of aircraft de-icing solutions on airport storm water runoff. They found that groundwater in the perched water table of sandy soil aquifer at the Ottawa International Airport (Canada) contained MPG at levels up to 4 mg/L in June, but declined to non-detectable levels by the fall. These findings verify the expected occurrence of MPG in surface run-off and groundwater that is associated with sites where de-icing and anti-icing formulations are applied. They also demonstrate the expected rapid dissipation (degradation) of the MPG substance, when emission to the surface water and groundwater environment is terminated.

Ongoing government-mandated environmental monitoring programs, which include the PG substances as target analytes, appear to be limited to a single program in Japan that is part of the Japan Ministry of Environment (MOE) environmental survey program for high-production and priority pollutant substances. In 1977 and 1986, the MOE surveyed water, bottom sediment, fish, and air samples collected from around the country for the presence of MPG (Ministry of Environment, Japan 2013). The MPG substance was not detected in any of the six surface water and six sediment samples collected in 1977. During a similar sampling of surface water and sediments in 1986, MPG was detected in 12 of 24 surface water samples, with detected concentrations ranging between 0.2 and 0.8 µg/L. Similarly, MPG was detected in 4 of 24 sediment samples, with concentrations ranging between 0.020 and 0.022 µg/g dry wt. Environmental monitoring data are not reported for MPG, or for any of the other PG substances beyond the 1986 campaign, which would indicate that the substances were identified as, and now remain as, low priorities for further investigation.

The Substances in Preparations in Nordic Countries (SPIN) database provides a qualitative assessment of consumer and environmental exposure potentials for chemicals used in consumer products in Norway, Sweden, and Denmark (http://www.SPIN200.net). The database indicates that one or more known product uses present the potential for "very probable" exposures of MPG, DPG, and TPG to air, water, soil, and wastewater media. The TePG substance is indicated as having one or several uses, with which only a "low" potential for exposure to wastewater is associated.

An example of the most extensive and relevant environmental monitoring was performed for PG substances by the U.S. EPA, which was associated with application of crude oil dispersants to remediate the 2010 Gulf of Mexico (Deepwater Horizon) oil spill. During the spill response (May to July 2010), an estimated total of 1.84 million gallons of dispersants, including COREXIT® EC9500A,[1] were applied both at the surface and directly at the wellhead on the seafloor (OSAT 2010). One of the ingredients in COREXIT® EC9500 is MPG, which comprises 1–5% (wt/wt) of the dispersant (Nalco Company 2008). It is estimated that approximately 0.7 million pounds of MPG was applied to the Gulf of Mexico oil spill response area. Between early May 2010 and late October 2010, over 17,000 samples were collected of water and sediment in the Gulf of Mexico area to locate oil and/or dispersant-related chemicals associated with the oil spill. Of all samples collected, only six sediment (0.73–1.0 µg/g) and two water samples (590 and 660 µg/L) contained MPG above the method detection limits of 0.5 µg/g and 500 µg/L, respectively (OSAT 2010).

---

[1] COREXIT® is a registered trademark of Nalco Company

These environmental monitoring programs and associated data, although limited in number and in geographic/temporal scope, indicate that despite the enormous tonnages of MPG used in numerous dispersive applications, the resultant concentration of MPG in environmental media is very low and usually non-detectable. Although very little or no environmental monitoring data are available for DPG, TPG or TePG, the Level III fugacity model results (as illustrated in Table 4b–d) demonstrate that these PG substances would have similar environmental distribution patterns to that of MPG (Table 4a) if they were used in similar amounts and modes of emission. Expected concentrations of these other PG substances in the environment would be even lower than observed or expected for MPG, because they are manufactured and used in lesser tonnages.

4 Environmental Fate Processes

The key processes that affect the persistence of substances in the atmospheric, aquatic, and terrestrial environments include photolysis (both direct and indirect), hydrolysis, and biodegradation. Other fate processes such as adsorption, volatilization, and bioaccumulation can affect the distribution and transport of substances within and among these environmental compartments. The relevance of these fate processes to the PG substances, along with summaries of the rates and extents to which the relevant processes occur, are discussed below.

4.1 Atmospheric Fate/Transport

The vapor pressures and Henry's Law Constants of the PG substances would not indicate significant prospective volatilization of the substances to the atmosphere. However, processing or use of them at elevated temperature, or the use and emission of their formulations directly in the atmosphere (as with aircraft de-icing and anti-icing formulations) could introduce them intermittently to the troposphere. As is illustrated by simulated atmospheric emissions using the Level III fugacity model (Table 4a–d), the fate of the PG substances in the atmospheric environment is governed by a combination of reactive, advective, and depositional processes.

The direct photolysis rate of substances in the atmosphere is governed by the band of wavelengths over which a particular molecule will absorb relevant solar radiation, the probability of a chemical reaction occurring per unit of photons absorbed (*i.e.*, quantum yield), and the intensity (*i.e.*, solar flux) at the relevant wavelengths of absorption. For the PG substances, the UV/VIS absorbance spectra each indicate a minor UV absorbance band over approximately 250–300 nm (data not shown). However, the wavelength band of sunlight that reaches the earth's surface is significantly filtered by ozone, water vapor, etc. in the upper atmosphere, such that irradiation by solar UV light is essentially cut off below 290 nm.

For this reason, direct photolysis of the PG substances is an unimportant fate process in the tropospheric, aquatic, and terrestrial environments.

As for most organic chemicals, the dominant reactive fate process for PG substances in the troposphere is indirect photolytic reaction with photochemically-produced hydroxyl radicals. During daylight hours, sunlight of wavelength <230 nm is absorbed by ozone in the troposphere, and forms a reactive atomic oxygen species. This reactive oxygen radical then reacts with atmospheric water to form highly reactive OH radicals. The reaction of PG substances with OH radical in the vapor phase occurs *via* hydrogen abstraction from the aliphatic –CH, –CH$_2$, and –CH$_3$ groups, and *via* reaction with the primary and secondary –OH groups. The products of OH radical reaction with these functional groups are expected to be various mono- and poly-carboxylates (aldehyde, ketone, and carboxylic acids), and ultimately CO$_2$.

The kinetics for reaction of MPG vapor with photochemically-generated OH radicals have been evaluated and reported by Atkinson (1986). A second-order reaction rate constant of 1.2×10^{-11} cm^3/molecule*s is reported for a temperature of 25 °C, and as shown in Table 3, is in excellent agreement with the estimated value from the AOPWIN software v1.92a (USEPA 2012). The second-order reaction rate constants for the DPG, TPG, and TePG substances and their associated structural isomers are also summarized in Table 3. Note that differences in atom connectivity among constitutional isomers of DPG, TPG, and TePG substances do not translate to significant differences in predicted rate constants for their reaction with OH radical. These rate constants equate to estimated atmospheric half-lives ranging from 1.6 to 10 h, at an assumed background hydroxyl radical concentration of 1.5×10^6 molecules/cm^3 and temperature of 25 °C. Substances that have tropospheric half-lives of >2 days are considered to be persistent in the environment by some regulatory authorities, and have potential for long-range transport *via* atmospheric advection (Calamari et al. 2000). Based on these estimated half-lives for indirect photolysis, it is concluded that the PG substances are rapidly degraded when emitted to the atmosphere, and have virtually no potential for long-range transport therein.

4.2 Biodegradation

Biodegradation is one of the most important processes influencing the persistence of organic chemicals in the environment. Several researchers, as described below, have evaluated the biodegradation of various PG substances using various inoculum sources or densities, substrate concentrations, and incubation conditions (Table 5). The biodegradation of these substances has been recently and thoroughly evaluated, using current OECD guidelines for testing of ready biodegradability, and biodegradability in seawater (West et al. 2007). The publication of these results included an in-depth review of current knowledge on metabolic pathways of their biodegradation, and on the physical-chemical and structural features that influence biodegradability. As noted above for the physico-chemical properties of these substances, numerous screening tests of ready and inherent biodegradability have been conducted over several decades. However, in many cases the important details on identity/purity of the tested substances, as well as those on experimental methods and inocula employed are lacking. The most recent results reported by West et al. (2007) are based on current standardized test methods, were conducted in accordance with GLP guidelines, and utilized thoroughly documented test substances and experimental procedures. Hence, they provide a definitive and reliable basis for assessing the ready biodegradability, biodegradation in seawater, and structure-biodegradability relationships across this family of substances. The results of these studies showed that six of the tested substances (MPG, DPG, TPG, PPG 425, PPG 1000, and PPG 2000) were readily biodegradable, whereas TePG and PPG 2700 were not readily biodegradable, but were inherently biodegradable. Biodegradation half-lives for these eight substances ranged from 3.8 days (PPG 2000) to 33.2 days (PPG 2700) in the ready test, and from 13.6 days (MPG) to 410 days (PPG 2700) in seawater tests. A further compilation of historical test results relating to these parameters is not presented here. Rather, results of selected biodegradation studies conducted in specific aquatic and terrestrial environments, and under aerobic and anaerobic conditions, are summarized below and in Table 5.

MPG has been shown to readily biodegrade in various screening tests employing non-adapted wastewater inocula under aerobic conditions (Kaplan et al. 1982; Price et al. 1974), and like many synthetic organic substances, is more rapidly biodegraded in acclimated systems in which bacteria with prior exposure and adapted metabolic systems exist (OECD 2001a). Kaplan et al. (1982) also demonstrated that MPG disappeared after 9 days under anaerobic conditions, when used as the sole carbon source by sludge from a sewage treatment plant. In simulation tests employing river waters, MPG was found to biodegrade rapidly as well (Gotvajn and Zagorc-Koncan 1999). Complete biodegradation of DPG and TPG was observed in the OECD 302B test of inherent biodegradability (OECD 1994, 2001b) and in the OECD 301E test of ready biodegradability (Zgola-Grzeskowiak et al. 2008); however, <3% biodegradation was observed for both DPG and TPG in the OECD 301C test for ready biodegradability (MITI 1995). This apparent lack of biodegradability in the OECD 301C test is believed to be associated with culturing of the inoculum on glucose and peptone, as discussed by West et al. (2007).

The biodegradation of PPGs has not been extensively studied, and while not directly within the scope of this review, it is worth noting that the rapid and complete biodegradation observed for the oligomeric PG substances is carried through to the polymeric PPG substances having molecular weight of up to ~2,000 g/mol (West et al. 2007). More recent studies showed complete biodegradation of PPG 725 and 40% biodegradation of PPG 425 in the OECD 301E test (Zgola-Grzeskowiak et al. 2007), whereas another study showed primary biodegradation of PPG 425 to an extent of 99% in a 17 days simulation test employing river water (Zgola-Grzeskowiak et al. 2006).

Besides biodegradation in the aquatic environment, the PG substances have also been observed to biodegrade in soil. Fincher and Payne (1962) and Kawai (1987)

Table 5 Summary of biodegradation studies for the propylene glycol substances

Substance	Study type		Result	Reference
MPG	Screening	Ready biodegradability	Aerobic	79% degraded over 20 days (readily biodegradable)	Price et al. (1974)
MPG	Screening	Ready biodegradability	Aerobic	107% degraded over 28 days (readily biodegradable)	West et al. (2007)
MPG	Screening	Inherent biodegradability	Aerobic	100% removal after 4 days	Kaplan et al. (1982)
MPG	Screening	Inherent biodegradability	Anaerobic	100% removal after 9 days	Kaplan et al. (1982)
MPG	Screening	Inherent biodegradability	Aerobic	84–99% removal in 20–24 h	OECD (2001a)
MPG	Simulation	River water	Aerobic	87–100% Removal in 28 days	Gotvajn and Zagorc-Koncan (1999)
MPG	Simulation	Seawater	Aerobic	91–96% removal in 64 days	West et al. (2007)
MPG	Simulation	Soil	Aerobic	100% in 12 days	Klecka et al. (1993)
MPG	Simulation	Soil	Anaerobic	Degraded to methane	OECD (2001a)
MPG	Simulation	Anaerobic Digestor	Anaerobic	Degraded to methane	Sezgin and Tomuk (2013)
DPG	Screening	Ready biodegradability	Aerobic	<3% removal in 28 days	OECD (2001b)
DPG	Screening	Ready biodegradability	Aerobic	84.4% degraded over 28 days (readily biodegradable)	West et al. (2007)
DPG	Screening	Inherent biodegradability	Aerobic	100% removal in 28 days	OECD (2001b)
DPG	Simulation	Seawater	Aerobic	17–24% removal in 64 days	West et al. (2007)
TPG	Screening	Ready biodegradability	Aerobic	<3% removal in 28 days	OECD (1994)
TPG	Screening	Ready biodegradability	Aerobic	81.9% degraded over 28 days (readily biodegradable)	West et al. (2007)
TPG	Screening	Inherent biodegradability	Aerobic	100% removal in 28 days	Zgola-Grzeskowiak, et al. (2008)
TPG	Simulation	Seawater	Aerobic	34–46% removal in 64 days	West et al. (2007)

TePG	Screening	Ready biodegradability	Aerobic	42–52% removal in 28 days	West et al. (2007)
TePG	Simulation	Seawater	Aerobic	19–31% removal in 64 days	West et al. (2007)
PPG 425	Screening	Ready biodegradability	Aerobic	88.6% degraded over 28 days (readily biodegradable)	West et al. (2007)
PPG 425	Screening	Inherent biodegradability	Aerobic	40% removal in 28 days	Zgola-Grzeskowiak et al. (2007)
PPG 425	Simulation	River water	Aerobic	99% removal in 17 days	Zgola-Grzeskowiak et al. (2006)
PPG 425	Simulation	Seawater	Aerobic	42–57% removal in 64 days	West et al. (2007)
PPG 725	Screening	Inherent biodegradability	Aerobic	100% removal in 28 days	Zgola-Grzeskowiak et al. (2007)
PPG 1000	Screening	Ready biodegradability	Aerobic	93.6% degraded over 28 days (readily biodegradable)	West et al. (2007)
PPG 1000	Simulation	Seawater	Aerobic	33–45% removal in 64 days	West et al. (2007)
PPG 2000	Screening	Ready biodegradability	Aerobic	105% degraded over 28 days (readily biodegradable)	West et al. (2007)
PPG 2000	Simulation	Seawater	Aerobic	26–38% removal in 64 days	West et al. (2007)
PPG 2700	Screening	Ready biodegradability	Aerobic	32–33% removal in 28 days	West et al. (2007)
PPG 2700	Simulation	Seawater	Aerobic	6% removal in 64 days	West et al. (2007)

isolated soil bacteria that were capable of using MPG and DPG as sole carbon sources. Kawai (1987) also showed that such isolates could utilize PPG substances up to PPG 3000. The soil microbe *C. glycolicum* was demonstrated to degrade MPG under anaerobic conditions to acid and alcohol end products (Gaston and Stadtman 1963). *Desulfovibrio*, a sulfate-reducing bacterium isolated from anoxic soil of a rice field, was reported to degrade MPG to acetate in the presence of sulfate with the production of carbon dioxide (Ouattara et al. 1992). Sezgin and Tomuk (2013) studied the applicability of semi-continuous anaerobic (methanogenic) bioreactors to treat MPG wastewaters, such as are generated from surface runoff of aircraft deicer/anti-icer formulations. They demonstrated essentially 100% removal of chemical oxygen demand (COD) as MPG at reactor feed rates of up to 750 mg/m^3/day and sludge age of 20 days. In simulation tests employing soil, MPG was also degraded under both aerobic (Klecka et al. 1993) and anaerobic (OECD 2001a) conditions. Klecka et al. (1993) concluded that the factors influencing the rates of biodegradation of MPG in soils were substrate concentrations, soil types, and ambient soil temperatures: lower glycol concentrations, higher soil organic carbon content, and higher ambient soil temperatures (in the range of −2 to 25 °C) resulted in faster degradation of MPG in soil. The biodegradation rate of MPG in soil was reported to be 2.3 mg/kg soil/day at −2 °C, 27.0 mg/kg soil/day at 8 °C, and 93.3 mg/kg soil/day at 25 °C (Klecka et al. 1993). The ease with which MPG is biodegraded in soil and groundwater, combined with the efficient production of hydrogen during its biodegradation by anaerobic bacteria, has resulted in its growing application to bioremediation of soil and groundwater contaminants (Adrian and Arnett 2007; Jaesche et al. 2006; Jin et al. 2002; Klecka 1996).

According to the studies presented here and elsewhere, the PG substances can be characterized as being rapidly biodegradable by a wide variety of inocula under a wide variety of incubation conditions. It would appear that the biodegradation of the PG substances involves enzymes that possess low specificity and/or high functional redundancy, and the same biodegradation pathways may be operative across this entire family of substances. Mechanisms, or even microorganisms, involved in biodegradation of PG substances might not be highly specialized, and appear to be widespread in the environment. Therefore, the PG substances are expected to rapidly degrade in a variety of environments and have low potential to be persistent in aquatic, terrestrial, and benthic environments.

4.3 Hydrolysis

The molecular structures of the PG substances consist exclusively of aliphatic –C–C–, C–H, –C–O–(ether, alcohol) and OH bonds. None of these molecular bonds are known or expected to be susceptible to hydrolysis under the temperature and pH conditions that are of physiological or environmental relevance. Generally, the aliphatic glycols and associated glycol ethers are regarded as being highly resistant to hydrolysis; however, no definitive study was identified in which this lack of reactivity for the PG substances was evaluated. The SIAR report for TPG makes reference to a 1993 unpublished study conducted by the Japan Chemicals Inspection and Testing Institute (CITI) according to OECD Guideline 111, wherein the substance was shown to be stable at pH 4, 7, and 9 at 25 °C (OECD 1994). For the purpose of demonstrating this expected lack of hydrolytic reactivity for various product regulatory assessments of structurally-related substances, the hydrolysis of a representative glycol ether (*i.e.*, dipropylene glycol n-propyl ether) has been evaluated as a function of pH, according to OECD Guideline 111: Hydrolysis as a Function of pH (ECHA 2013b). The substance tested possesses all of the same structural features and molecular bonds that are represented across the PG substances. In this study, no degradation of the substance was observed over a 5-days exposure to pH 7 and 9 buffer solutions at 50 °C. Less than 4% degradation was observed under the same conditions at pH 4, and the substance was concluded to be hydrolytically stable. The half-life for hydrolysis of this tested representative substance, and for any of the PG substances by analogy, can be expected to exceed 1 year at 25 °C exposure, within the pH range of 4–9. Thus, hydrolysis is confirmed to be an unimportant fate process for the PG substances.

4.4 Bioaccumulation

Considering their miscibility with water, very low log P_{ow} values, and ability to be readily metabolized in microorganisms and in higher animals, the PG substances are expected to exhibit very low or no potential to bioaccumulate in the aquatic environment, or to biomagnify in the food chain of terrestrial vertebrates. Despite low bioaccumulation potential, the bioconcentration of the DPG and TPG substances have been evaluated in fish, according to OECD Guideline 305: Flow-through test (MITI 1995). The measured fish bioconcentration factor (BCF) for DPG in *Cyprinus carpio* ranged from 0.3 to 4.6 L/kg, and that for TPG in the same species was not measurable (BCF <5.7 L/kg). Propylene glycol substances of higher molecular weight would appear to have the same low potential to bioaccumulate. A PPG substance having a molecular weight of 3,000 g/mol was associated with measured fish BCF values of <7 and <2.2 L/kg, using the same species and similar test procedures to those used for DPG and TPG (CERI 1977). These measured BCF values are consistent with estimated BCF values produced by the US EPA BCFBAF model (v3.01, USEPA 2012), which are based on correlation of BCF with log P_{ow}. Using the log P_{ow} values shown in Table 2, the same BCF value of 3.16 L/kg is estimated for MPG, DPG, TPG and TePG. This BCF value of 3.16 L/kg is the *de minimus* BCF value reported by the BCFBAF model, for substances having log P_{ow} values of <1.0. Considering their physico-chemical properties and rapid degradability, along with measured fish BCF values for representative substances, it is concluded that the PG substances have very low potential to bioaccumulate in aquatic and terrestrial organisms.

5 Ecotoxicity

The PG substances consist of simple molecular structures that are not ionizable, and do not react directly with proteins or other cellular components of tissues. As such, any toxic effects resulting from either acute or chronic exposures to the substances at realistic concentrations would occur *via* a non-specific mode of action referred to as "non-polar narcosis". This minimum or base-line toxicity of substances is highly-correlated with hydrophobicity (*i.e.*, log P_{ow}) of substances, and can be thought of as the minimum degree of toxic potential likely to be exerted by any organic substance. Substances that exert toxic effects at lower concentrations than predicted from this base-line correlation with log P_{ow} are likely acting *via* one or more specific (*i.e.*, reactive) modes of action in parallel with narcosis (Roberts and Costello 2003). In the following sections, an overview of available acute and chronic studies with both aquatic and terrestrial organisms is presented, which exemplify the base-line toxicity exhibited by the PG substances.

5.1 Monopropylene Glycol (MPG)

The acute toxicity of MPG toward aquatic and terrestrial species has been well-studied across vertebrate, invertebrate, and plant species associated with both aquatic and terrestrial environments. As can be seen in Table 6a, the acute LC_{50} values for MPG exposures of all fish species tested are >1,000 mg/L. The LC_{50} and EC_{50} values associated with acute MPG exposures to clawed frog, all aquatic invertebrates and algae species tested, and lettuce are >10,000 mg/L. Overall, MPG is practically non-toxic to aquatic and terrestrial organisms on an acute basis.

Chronic exposure assays of MPG were also conducted with several species of aquatic and terrestrial organisms, and low potential for long-term adverse effects was exhibited. As shown in Table 6a, the 7-days chronic NOEC to fathead minnow and water flea, the 14-days EC_{50} to algae, and the 5-days EC_{25} for lettuce, are all >10,000 mg/L. The above data demonstrate that MPG has a very low order of toxicity in the aquatic and terrestrial environments.

5.2 Dipropylene Glycol (DPG)

The acute toxicity of DPG to several aquatic species including fish, frog, water flea, and algae has been determined. No acute toxicity data are available for DPG in terrestrial organisms. As can be seen in Table 6b, the LC_{50} values determined for DPG with fish and frog species tested are all >1,000 mg/L. The EC_{50} values for DPG exposures to the water flea and algae are all >100 mg/L. Based on the available data, DPG is also demonstrated to have a very low order of toxicity in the environment.

Table 6a Summary of aquatic and terrestrial toxicity data for monopropylene glycol (MPG)

Species	Endpoint and duration	Result	Reference
Aquatic vertebrates			
Goldfish *Carassius auratus*	24-h LC_{50}	>5,000 mg/L	Bridie et al. (1979)
Sheepshead minnow *Cyprinodon variegatus*	24-h LC_{50}	63,500 mg/L	USEPA (2000)
Sheepshead minnow *Cyprinodon variegatus*	48-h LC_{50}	52,500 mg/L	USEPA (2000)
Sheepshead minnow *Cyprinodon variegatus*	72-h LC_{50}	35,900 mg/L	USEPA (2000)
Sheepshead minnow *Cyprinodon variegatus*	96-h LC_{50}	23,800 mg/L	USEPA (2000)
Sheepshead minnow *Cyprinodon variegatus*	96-h LC_{50}	48,000 mg/L	Mayer and Ellersieck (1986)
Guppy *Lebistes reticulatus*	48-h LC_{50}	>10,000 mg/L	Verschueren (2001)
Bluegill sunfish *Lepomis macrochirus*	96-h LC_{50}	>10,000 mg/L	USEPA (2006)
Inland Silverside *Menidia beryllina*	96-h LC_{50}	>10,000 mg/L	USEPA (2006)
Rainbow trout *Oncorhynchus mykiss*	24-h LC_{50}	79,700 mg/L	USEPA (2000)
Rainbow trout *Oncorhynchus mykiss*	24-h LC_{50}	50,000 mg/L	Verschueren (2001)
Rainbow trout *Oncorhynchus mykiss*	48-h LC_{50}	79,700 mg/L	USEPA (2000)
Rainbow trout *Oncorhynchus mykiss*	72-h LC_{50}	51,600 mg/L	USEPA (2000)
Rainbow trout *Oncorhynchus mykiss*	96-h LC_{50}	51,600 mg/L	USEPA (2000)
Rainbow trout *Oncorhynchus mykiss*	96-h LC_{50}	44,000 ppm	Mayer and Ellersieck (1986)
Rainbow trout *Oncorhynchus mykiss*	96-h LC_{50}	42,380 and 37,067 mg/L	USEPA (2000)
Rainbow trout *Oncorhynchus mykiss*	96-h LC_{50}	45,600 mg/L	Mayer and Ellersieck (1986)
Medaka *Oryzias latipes*	48-h LC_{50}	>1,000 mg/L (static)	Tsuji et al. (1986)
Fathead minnow *Pimephales promelas*	24-h LC_{50}	77,800 mg/L	USEPA (2000)
Fathead minnow *Pimephales promelas*	48-h LC_{50}	54,000 mg/L	USEPA (2000)
Fathead minnow *Pimephales promelas*	72-h LC_{50}	51,400 mg/L	USEPA (2000)
Fathead minnow *Pimephales promelas*	96-h LC_{50}	51,400 mg/L	USEPA (2000)
Fathead minnow *Pimephales promelas*	96-h LC_{50}	59,900–77,400 mg/L	USEPA (2006)

(continued)

Table 6a (continued)

Species	Endpoint and duration	Result	Reference
Fathead minnow *Pimephales promelas*	96-h LC_{50}	54,900 mg/L	Verschueren (2001)
Fathead minnow *Pimephales promelas*	96-h LC_{50}	34,060 mg/L	Cornell et al. (2000)
Fathead minnow *Pimephales promelas*	96-h LC_{50}	55,770 mg/L NOEC mortality = 52,930	Pillard (1995)
Fathead minnow *Pimephales promelas*	7-days NOEC growth and mortality	<11,530 mg/L	Pillard (1995)
Fingerling trout *Salmo gairdneri*	24-h LC_{50}	50,000 mg/L	Majewski et al. (1978)
Clawed Frog *Xenopus laevis*	48-h LC_{50}	18,700 and 24,285 mg/L	USEPA (2000)
Aquatic invertebrates			
Water flea *Ceriodaphnia dubia*	48-h LC_{50}	18,340 mg/L NOEC = 13,020 mg/L	Pillard (1995)
Water flea *Ceriodaphnia dubia*	7-days NOEC	13,020 mg/L (reproduction) 29,000 mg/L (mortality)	Pillard (1995)
Water flea *Daphnia magna*	24-h LC_{50}	70,700 mg/L	USEPA (2000)
Water flea *Daphnia magna*	24-h EC_{50} immobilization	>10,000 mg/L	Kuhn et al. (1989)
Water flea *Daphnia magna*	48-h LC_{50}	43,500 mg/L	USEPA (2000)
Brine Shrimp *Artemia salina*	24-h LC_{50}	>10,000 mg/L	Price et al. (1974)
Mysid shrimp *Mysidopsis bahia*	24-h LC_{50}	31,000 mg/L	USEPA (2000)
Mysid shrimp *Mysidopsis bahia*	48-h LC_{50}	27,300 mg/L	USEPA (2000)
Mysid shrimp *Mysidopsis bahia*	72-h LC_{50}	23,400 mg/L	USEPA (2000)
Mysid shrimp *Mysidopsis bahia*	96-h LC_{50}	18,800 mg/L	USEPA (2000)
Mysid shrimp *Mysidopsis bahia*	96-h LC_{50}	11,000 ppm	Mayer and Ellersieck (1986)
Harpacticoid copepod *Nitocra spinipes*	96-h LC_{50}	>10,000 mg/L	Tarkpea et al. (1986)
Green algae *Selenastrum capricornutum*	48-h EC_{50} Growth rate	34,100 mg/L	USEPA (2000)
Green algae *Selenastrum capricornutum*	72-h EC_{50} Growth rate	24,200 mg/L	USEPA (2000)
Green algae *Selenastrum capricornutum*	96-h EC_{50} Growth rate	19,000 mg/L	USEPA (2000)

(continued)

Table 6a (continued)

Species	Endpoint and duration	Result	Reference
Green algae *Selenastrum capricornutum*	96-h	IC_{50} = 20,690 mg/L IC_{25} = 1,516 mg/L LOEC = 126 mg/L NOEC = 37 mg/L	USEPA (2000)
Green algae *Selenastrum capricornutum*	96-h IC_{25}	20,800 mg/L	USEPA (2000)
Green algae *Selenastrum capricornutum*	14-days EC_{50} Growth rate	18,100 mg/L	USEPA (2000)
Marine algae *Skeletonema costatum*	24-h EC_{50} Growth rate	31,500 mg/L	USEPA (2000)
Marine algae *Skeletonema costatum*	48-h EC_{50} Growth rate	19,000 mg/L	USEPA (2000)
Marine algae *Skeletonema costatum*	72-h EC_{50} Growth rate	19,300 mg/L	USEPA (2000)
Marine algae *Skeletonema costatum*	96-h EC_{50} Growth rate	19,100 mg/L	USEPA (2000)
Marine algae *Skeletonema costatum*	14-days EC_{50} Growth rate	<5,300 mg/L	USEPA (2000)
Duckweed *Lemna minor*	96-h	IC_{25} = 12,000 mg/L (frond growth) LOEC = 5,000 mg/L (frond growth) IC_{25} = 21,882 mg/L (chlorophyll) LOEC = 20,000 mg/L (chlorophyll) IC_{25} = 12,000 mg/L (pheophytin) LOEC = 20,000 mg/L (pheophytin)	USEPA (2000)
Toxicity to terrestrial plants			
Lettuce *Lactuca sativa*	72-h EC_{50} Germination	50,540 mg/L	Reynolds (1977)
Lettuce *Lactuca sativa*	5-days EC_{25} (hydroponic)	24,760 mg/L (emergence) NOEC = 4,500 mg/L 9,880 mg/L (root length) 1,190 mg/L (shoot length)	Pillard and Dufresne (1999)
Ryegrass *Lolium perenne*	5-days EC_{25} (hydroponic)	24,210 mg/L (emergence) NOEC = 15,000 mg/L 2,850 mg/L (root length) 3,120 mg/L (shoot length)	Pillard and Dufresne (1999)
Toxicity to other non-mammalian terrestrial species (including birds)			
Domestic Chicken **embryo** *Gallus domesticus*	14-days NOEL (chick embryo mortality)	0.05 ml/embryo	Gebhardt and Van Logten (1968)

Table 6b Summary of aquatic and terrestrial toxicity data for dipropylene glycol (DPG)

Species	Endpoint and duration	Result	Reference
Aquatic vertebrates			
Goldfish	24-h LC_{50}	>5,000 mg/L	Bridie et al. (1979)
Carassius auratus			
Clawed Frog	48-h LC_{50}	3,181 mg/L	De Zwart and
Xenopus laevis			Slooff (1987)
Aquatic invertebrates			
Water flea	48-h EC_{50} immobilization	>100 mg/L	ECHA (2013c)
Daphnia magna			
Aquatic plants			
Algae	72-h EC_{50}	>100 mg/L	ECHA (2013c)
Desmodesmus subspicatus	Growth inhibition	NOEC >100 mg/L	

5.3 Tripropylene Glycol (TPG)

TPG has been tested in a limited number of aquatic species for acute and chronic toxicity. No toxicity data are available for TPG in terrestrial organisms. As can be seen in Table 6c, the available LC_{50}/EC_{50} values associated with acute exposures of TPG to fish, water flea, and algae are all >1,000 mg/L. The chronic 21-day NOEC (reproduction and immobility) for water flea is also >1,000 mg/L. Therefore, TPG is considered to be practically non-toxic to fish, daphnids, and algae, and it does not have any remarkable ecotoxicity.

5.4 Tetrapropylene Glycol (TePG) and Higher Oligomers

Due to overlap in their molecular weight, ecotoxicity information for the TePG substance (M_n = 250 g/mol) is discussed along with that for low molecular weight PPG substances. Limited acute toxicity data are available for PPG exposures to aquatic organisms. No acute toxicity data are available for terrestrial organisms, and no chronic toxicity data are available for either aquatic or terrestrial organisms. As can be seen in Table 6d, the LC_{50}/EC_{50} values associated with acute exposures of PPG (M_n = 260 g/mol) to fish, water flea and algae are all >100 mg/L. A 3-h EC_{50} >1,000 mg/L of PPG (M_n = 230 g/mol) was also reported for bacterial growth inhibition. These results suggest a very low toxicity of PPGs in the environment.

It is concluded from the available data summarized here that the PG substances do not pose short- or long-term risks to environmental receptors at concentrations that could reasonably be expected to result from typical use and disposal patterns. In standardized tests of acute aquatic toxicity, the maximum recommended (limit) exposure concentration is typically 100 mg/L. According to regulatory classification schemes for acute aquatic toxicity, substances exhibiting E/LC_{50} values of >100 mg/L are regarded as "practically non-toxic" and are not classified for acute toxic effects. It is therefore important to note that none of the acute tests summarized here for the PG substances resulted in E/LC_{50} values of <100 mg/L.

Table 6c Summary of aquatic and terrestrial toxicity data for tripropylene glycol (TPG)

Species	Endpoint and duration	Result	Reference
Aquatic vertebrates			
Medaka *Oryzias latipes*	96-h LC_{50}	>1,000 mg/L (semi-static)	Environment Agency Japan (1992)
Common carp *Cyprinus carpio*	Bioaccumulation (OECD 305)	BCF: <5.7 (1 mg/L) BCF: <0.5 (10 mg/L)	MITI (1995)
Aquatic invertebrates			
Water flea *Daphnia magna*	24-h EC_{50} immobilization	>1,000 mg/L (static)	Environment Agency Japan (1992)
Water flea *Daphnia magna*	21-day NOEC Reproduction and immobility	>1,000 mg/L (semi-static)	Environment Agency Japan (1992)
Aquatic plants			
Green algae *Pseudokirchnerella subcapitata* (reported as *Selenastrum capricornutum*)	72-h EC_{50} Biomass growth inhibition	>1,000 mg/L NOEC >1,000 mg/L	Environment Agency Japan (1992)

Table 6d Summary of aquatic and terrestrial toxicity data for tetrapropylene glycol (TePG)

Species	Endpoint and duration	Result	Ave. MW (g/mol)	Reference
Aquatic vertebrates				
Zebrafish *Danio rerio*	96-h LC_{50}	>100 mg/L (static)	260	ECHA (2013d)
Aquatic invertebrates				
Water flea *Daphnia magna*	48-h EC_{50}	105.8 mg/L (static)	260	ECHA (2013d)
Aquatic plants				
Algae *Desmodesmus subspicatus*	72-h EC_{50} (growth rate)	>100 mg/L (static) NOEC = 100 mg/L	260	ECHA (2013d)
Microorganisms				
Activated sludge	3-h EC_{50} (respiration rate)	>1,000 mg/L NOEC = 1,000 mg/L	230	ECHA (2013d)

Similarly, substances exhibiting chronic NOEC or EC_{10} values of >1 mg/L are typically not classified as having potential to cause long-term effects in the environment. Although it might be of interest to examine the potential correlation of acute and chronic effect levels with log P_{ow} values of the PG substances, testing in most cases involved limit concentration exposures (*i.e.*, 100; 1,000; 10,000 mg/L), from which discrete values of E/LC_{50} and NOEC were not determinable (Tables 6a, 6b, 6c, and 6d). Therefore, it is not possible to determine an approximate ratio of acute:chronic toxicity threshold concentrations from the available ecotoxicological datasets for these substances. Even in the absence of these refined analyses of their toxicity potentials, the empirical data on these substances clearly indicate that acute and chronic effects are not expected to occur for typical and recommended use and disposal of the products containing them. This, combined with demonstrated rapid and ultimate biodegradability and lack of bioaccumulation potential, leads to the conclusion that the PG substances have low potential for environmental harm.

6 Potential for Endocrine Disruption

The potential for xenobiotic substances to interfere with endocrine modulation in humans and wildlife is a topic of high current interest. As a result of concern for these potential effects from pesticides, persistent organic pollutants, and other substances produced in large volumes, regulatory authorities are requiring evaluations for endocrine disrupting potential of such substances. Searches of the published literature, government databases, and internet did not locate information pertaining to direct assessment or association of endocrine modulating effects for the PG substances. However, considering the widespread and often dispersive uses of these substances, along with the aforementioned sporadic detections of PG and DPG in surface-, ground-, and drinking-water samples, this review of the environmental fate and effects of these substances might be considered incomplete without presenting the following weight of indirect evidence regarding potential for endocrine effects of the PG substances.

MPG is considered by the U.S. Food and Drug Administration to be a Generally Recognized as Safe (GRAS) substance, and as reviewed recently by Fowles et al. (2013), has been extensively tested for potential effects on development and reproduction of mammals. These varied and numerous studies revealed no effects on mammalian reproductive performance, fetal development, or histopathological evidence of endocrine-mediated effects in reproductive toxicity studies with the PG substances. The Center for the Evaluation of Risks to Human Reproduction (CERHR), a division of the National Institute of Environmental Health Sciences (NIEHS), reviewed the potential reproduction/developmental effects of MPG in 2004, and concluded that the substance is "of negligible concern for reproduction/developmental effects" (CERHR 2004). The MPG substance is employed as an excipient in various oral and injectable therapies (both prescription and OTC/herbal) used to manage estrogen, androgen, and thyroid hormone levels in humans.

Finally, evidence from structure-activity relationships can be used to evaluate the affinity that the PG substances and their associated isomers have for binding to the estrogen and androgen receptors. The OASIS TIssue MEtabolism SImulator model (*i.e.*, OASIS TIMES v2.27.5, Laboratory of Mathematical Chemistry of the University of Professor Assen Zlatarov, Bourgas, Bulgaria) employing a heuristic probabilistic algorithm (Mekenyan et al. 2004), was used to estimate the estrogen- and androgen-receptor binding affinity for the PG substances. The major representative isomer for each glycol was used for dipropylene and higher PG oligomers. The TIMES modeling is based on a Common Reactivity Pattern (COREPA) approach which assesses the impact of three-dimensional molecular conformation distributions and flexibility on stereo-electronic properties of the modeled substances (Mekenyan and Serafimova 2009). The modeling predicted that each of the PG substances and their various associated 3-D molecular conformers would be "not active" with the human estrogen and androgen nuclear receptors. Thus, the modeling found that these substances have no potential for endocrine disruption *via* direct receptor binding agonist or antagonist modes of action. In more general terms, the overall chemical structures of the PG substances are not indicative of endocrine disruption properties, as these compounds lack certain structural features that appear to be important for nuclear binding affinity, such as hydrogen bond donor and acceptor groups associated with single or multiple aromatic rings. Based on the experimental findings and modeling results of receptor binding affinities, the PG substances are not considered to be potential endocrine disruptors, such that they could induce endocrine-modulating effects on humans, fish, or other wildlife.

7 Summary

The propylene glycol substances comprise a homologous family of synthetic organic molecules that have widespread use and very high production volumes across the globe. The information presented and summarized here is intended to provide an overview of the most current and reliable information available for assessing the potential environmental exposures and impacts of these substances across the manufacture, use, and disposal phases of their product life cycles.

The PG substances are characterized as being miscible in water, having very low octanol-water partition coefficients (log P_{ow}) and exhibiting low potential to volatilize from water or soil in both pure and dissolved forms. The combination of these properties dictates that, almost regardless of the mode of their initial emission, they will ultimately associate with surface water, soil, and the related groundwater compartments in the environment. These substances have low affinity for soil and sediment particles, and thus will remain mobile and bio-available within these media.

In the atmosphere, the PG substances are demonstrated to have short lifetimes (1.7–11 h), due to rapid reaction with photochemically-generated hydroxyl radicals. This reactivity, combined with efficient wet deposition of their vapor and aerosol forms, lends to their very low potential for long-range transport *via* the atmosphere. In the aquatic and terrestrial compartments of the environment, the PG substances are rapidly and ultimately biodegraded under both aerobic and anaerobic conditions by a wide variety of microorganisms, regardless of prior adaptation to the substances. Except for the TePG substance, the propylene glycol substances meet the OECD definition of "readily biodegradable", and according to this definition are not expected to persist in either aquatic or terrestrial environments. The TePG exhibits inherent biodegradability, is not regarded to be persistent, and is expected to ultimately biodegrade in the environment, albeit at a somewhat slower rate.

The apparent ease with which microorganisms and higher organisms can metabolize the PG substances, along with their low log P_{ow} and very high water solubility values, portends them to have very low potential for bioaccumulation and/or biomagnification in aquatic and terrestrial organisms. These same properties, along with their neutral structures and lack of biological reactivity, are the reasons for which the PG substances exhibit a base-line, non-polar narcosis mode of toxicity. The PG substances have been shown here to be practically non-toxic to essentially every aquatic and terrestrial animal and plant species tested. Collectively, the available wealth of information relating to persistence, bioaccumulation, and eco-toxicity of these substances allows a definitive conclusion of their categorization as not being PBT (*i.e.*, persistent/bioaccumulative/toxic). The PBT screening and categorization of substances on the Canadian Domestic Substances List (DSL) by Environment Canada has formally concluded that each member of this substance family is "not P", "not B", and "not T" according to their associated PBT criteria. Similarly, the preceding evaluations of these high production volume substances within the OECD SIDS program concluded that MPG, DPG, and TPG are low priorities for further examination of potential impacts to humans and the environment. More extensive evaluations of potential risks to human health and the environment were recently completed by industry, as required for their registration under the European Union REACh legislation; each evaluation demonstrated that current uses, associated exposures, and controls thereof, will not result in exposures that exceed predicted no effect concentrations in the environment.

Acknowledgements The authors wish to acknowledge the American Chemistry Council Propylene Oxide/Propylene Glycol Panel, and its Director, Mr. Jonathon Busch, for funding and administrative support for preparation of this review. All conclusions presented herein are those of the authors, and not necessarily of the American Chemistry Council or its sponsor companies.

References

Adrian NR, Arnett CM (2007) Anaerobic biotransformation of explosives in aquifer slurries amended with ethanol and propylene glycol. Chemosphere 66:1849–1856

AIChE (2012) DIPPR project 801. Design Institute for Physical Properties Research, American Institute of Chemical Engineers, New York, NY, Available at: http://www.aiche.org/dippr/projects/801

Atkinson R (1986) Kinetics and mechanisms of the gas-phase reactions of the hydroxyl radical with organic compounds under atmospheric conditions. Chem Rev 86:69–201

Bridie A, Wolff CJM, Winter M (1979) The acute toxicity of some petrochemicals to goldfish. Water Res 13:623–626

Brown ES, Hauser CF, Ream BC, Berthold RV (1980) Glycols. In: Kirk-Othmer encyclopedia of chemical technology, Band 11, 3.Aufl. Wiley, New York, NY, pp 933–956

Calamari D, Jones K, Kannon K, Lecloux A, Olsson M, Thurman M, Zannetti P (2000) Monitoring as an indicator of persistence and long-range transport. In: Klecka G, Boethling R, Franklin J, Grady L Jr, Graham D, Howard PH, Kannan K, Larson B, Mackay D, Muir D, van de Meent D (eds) Evaluation of persistence and long-range transport of organic chemicals in the environment. SETAC Press, Pensacola, FL, p 103

CEMC (2004) Level III fugacity-based multimedia environmental model, v2.80.1. Canadian Environmental Modelling Centre, Trent University, Peterborough, ON, Available from: htpp://trentu.ca/cemc/

CERHR (2004) Monograph on the potential human reproductive and developmental effects of propylene glycol. National Toxicology Program, Center for Evaluation of Risk to Human Reproduction. United States National Institutes of Health Publication No. 04-4482

CERI (1977) Bioconcentration study for propoxylated glycerol, reported in: Biodegradation and bioaccumulation data for existing chemicals. Chemicals Evaluation and Research Institute, Tokyo, Japan, Available at: http://www.safe.nite.go.jp/english/db.html

Chinn H, Kumamoto T (2011) Chemical economics handbook (CEH) product review: propylene glycols. SRI Consulting, Menlo Park, CA

Cornell JS, Pillard DA, Hernandez MT (2000) Comparative measures of the toxicity of component chemicals in aircraft deicing fluid. Environ Toxicol Chem 19:1465–1472

Cox RA, McCartney MJ, Culkin F (1970) The specific gravity/salinity/temperature relationship in natural sea water. Deep-Sea Res 17:679–689

De Zwart D, Slooff W (1987) Toxicity of mixtures of heavy metals and petrochemicals to *Xenopus laevis*. Bull Environ Contam Toxicol 38:345–351

Dimitrov S, Dimitrova N, Parkerton T, Comber M, Bonnell M, Mekenyan O (2005) Base-line model for identifying the bioaccumulation potential of chemicals. SAR QSAR Environ Res 16:531–554

Dunlap WJ, Shew DC (1976) Organic pollutants contributed to ground water by a landfill. U.S. Environmental Protection Agency, Report No. EPA-600/9-76-004. pp 96–110.

EC (2008) Council Regulation (EC) No 440/2008. Off J Eur Union L142:1–739

ECHA (2013a) Information on chemicals registered under REACh; 1,1,2,2-tetrachloroethylene, CAS Registry No. 127-18-4. European Chemicals Agency, Helsinki, Finland, Available at: http://echa.europa.eu/web/guest/information-on-chemicals/registered-substances

ECHA (2013b) Information on chemicals registered under REACh, 1-(1-methyl-2-propoxyethoxy) propan-2-ol, CAS Registry No. 29911-27-1. European Chemicals Agency, Helsinki, Finland, Available at: http://echa.europa.eu/web/guest/information-on-chemicals/registered-substances

ECHA (2013c) Information on chemicals registered under REACh, oxydipropanol, CAS Registry No. 25265-71-8. European Chemicals Agency, Helsinki, Finland, Available at: http://echa.europa.eu/web/guest/information-on-chemicals/registered-substances

ECHA (2013d) Information on chemicals registered under REACh, 1,2-propanediol, propoxylated, CAS Registry No. 25322-16-4. European Chemicals Agency, Helsinki, Finland, Available at: http://echa.europa.eu/web/guest/information-on-chemicals/registered-substances

Environment Agency Japan (1992) Investigation of the ecotoxicological effects of OECD high production volume chemicals. Office of Health Studies, Environmental Health Department, Environment Agency, Japan

Fincher EL, Payne WJ (1962) Bacterial utilization of ether glycols. Appl Microbiol 10:542–547

Fowles JR, Banton MI, Pottenger LH (2013) A toxicological review of the propylene glycols. Crit Rev Toxicol 43(4):363–390

Gaston LW, Stadtman ER (1963) Fermentation of ethylene glycol by *Clostridium glycolicum*. J Bacteriol 85:356–362

Gebhardt DOE, Van Logten MJ (1968) The chick embryo test as used in the study of the toxicity of certain dithiocarbamates. Toxicol Appl Pharmacol 13:316–324

Gotvajn AZ, Zagorc-Koncan J (1999) Laboratory simulation of biodegradation of chemicals in surface waters: closed bottle and respirometric test. Chemosphere 38:1339–1346

Jaesche P, Totsche KU, Kogel-Knabner I (2006) Transport and anaerobic biodegradation of propylene glycol in gravel-rich soil materials. J Contam Hydrol 85(3–4):271–286

Jin P, Droy BF, Manale F, Liu S, Copeland R, Creber C, Klecka G (2002) Monitoring the effectiveness of large-scale in situ anaerobic bioremediation. In: Remediation of chlorinated and recalcitrant compounds—2002. Proceedings of the international conference on remediation of chlorinated and recalcitrant compounds, 3rd, Monterey, CA, United States, 20–23 May. pp. 797–805

Kaplan DL, Walsh JT, Kaplan AM (1982) Gas chromatographic analysis of glycols to determine biodegradability. Environ Sci Technol 16:723–725

Kawai F (1987) The biochemistry of degradation of polyethers. Crit Rev Biotechnol 6:273–307

Klecka GM (1996) Method for stimulating anaerobic biotransformation of halogenated hydrocarbons. United States patent No. 5578210. The Dow Chemical Company, Midland, MI

Klecka GM, Carpenter C, Landenberger BD (1993) Biodegradation of aircraft deicing fluids in soil at low temperatures. Ecotoxicol Environ Saf 25:280–295

Kuhn R, Pattard M, Pernak K, Winter A (1989) Results of the harmful effects of selected water pollutants (anilines, phenols, aliphatic compounds) to Daphnia magna. Water Res 23:495–499

Landolt-Bornstein (1980) Numerical data and functional relationships in physics, chemistry, geophysics, and technology, vol IV, 6th edn. Springer, New York, NY, pp 101–102

Larson R, Forney L, Grady L Jr, Klecka G, Masunaga S, Peijneburg W, Wolfe L (2000) Quantitation of persistence in soil, water, and sediments. In: Klecka G, Boethling R, Franklin J, Grady L Jr, Graham D, Howard PH, Kannan K, Larson B, Mackay D, Muir D, van de Meent D (eds) Evaluation of persistence and long-range transport of organic chemicals in the environment. SETAC Press, Pensacola, FL, p 103

Lin DDD, Melton RG, Kopfler FC, Lucas SV (1981) Glass capillary gas chromatographic/mass spectrometric analyses of organic concentrates from drinking and advanced waste treatment water. In: Keith LH (ed) Advances in the identification and analysis of organic pollutants in water, vol 2. Ann Arbor Science Publishers, Ann Arbor, MI, pp 861–906

Lyman WJ (1990) Adsorption coefficient for soils and sediments. In: Lyman WJ, Reehl WF, Rosenblatt DH (eds) Handbook of chemical property estimation methods. American Chemical Society, Washington, DC

Mackay D (2001) Multimedia environmental models: the fugacity approach, 2nd edn. CRC Press LLC, Boca Raton, FL

Majewski H, Klaverkamp J, Scott D (1978) Acute mortality and sub-lethal effects of acetone, ethanol and propylene glycol on the cardiovascular and respiratory systems of rainbow trout (*Salmo gairdneri*). Water Res 13:217–221

Mayer FL, Ellersieck MR (1986) Manual of acute toxicity: interpretation and database for 410 chemicals and 66 species of freshwater animals. United States Fish and Wildlife Service, Washington, DC

Mekenyan O, Serafimova R (2009) Mechanism based modeling of estrogen receptor binding affinity: a common reactivity pattern (COREPA) implementation. In: Devillers J (ed) Endocrine disruption modeling. CRC Press LLC, Boca Raton, FL, pp 259–294

Mekenyan OG, Dimitrov SD, Pavlov TS, Veith GD (2004) A systematic approach to stimulating metabolism in computational toxicology. I. The TIMES heuristic modelling framework. Curr Pharm Des 10:1273–1293

Ministry of Environment Japan (2013) Repot on environmental survey and wildlife monitoring of chemicals. Chemical Risk Information Platform, Ministry of Environment, Japan, Available at: http://www.safe.nite.go.jp/english/db.html

MITI (1995) Biodegradation and bioconcentration of existing chemical substances under the chemical substances control law. Japan Chemical Industry Ecology Toxicology & Information Center, Japan

Moermond C, Janssen M, de Knecht J, Montforts M, Peijenburg W, Zweers P, Sijm D (2011) PBT assessment using the revised Annex XIII of REACH: a comparison with other regulatory frameworks. Integr Environ Assess Manag 8:359–371

Moran MJ, Delzer GC (2006) Contamination of ground water by PCE—a national perspective. In: Petroleum hydrocarbons and organic chemicals in ground water: prevention, assessment, and remediation conference, Houston, TX, 6–7 Nov 2006. United States Geological Survey (USGS), Publication No. 70033577. Available from: http://pubs.er.usgs.gov

Nalco Company (2008) Safety data sheet for product COREXIT (R) EC9500A. Nalco Company, Naperville, IL

OECD (1991) Second meeting of the OECD expert group on polymer definition: Chairman's report [ENV/MC/CHEM(91)18]. Organisation for Economic Co-operation and Development, Paris, France, October 1991

OECD (1994) SIDS initial assessment report for tripropylene glycol. UN Publication. Available at: http://www.chem.unep.ch/irptc/sids/OECDSIDS/sidspub.html

OECD (2001a) SIDS initial assessment report for 1,2-dihydroxypropane. UN Publication. Available at: http://www.chem.unep.ch/irptc/sids/OECDSIDS/sidspub.html

OECD (2001b) SIDS initial assessment report for dipropylene glycol. UN Publication. Available at: http://www.chem.unep.ch/irptc/sids/OECDSIDS/sidspub.html

OECD (2003) OECD principles on good laboratory practice, No. 1. Organisation for Economic Co-operation and Development, Paris, France. Revised 13 Feb 2003. Available at: http://www.oecd-ilibrary.org/

OECD (2012) OECD guidelines for the testing of chemicals—guideline, Section 1: Physical-chemical properties. Organisation for Economic Co-operation and Development, Paris, France. Updated 02 Oct 2012

OSAT (2010) Summary report for sub-sea and sub-surface oil and dispersant detection: sampling and monitoring. Operational Science Advisory Team, Unified Area Command. Available at: http://www.restorethegulf.gov

Ouattara AS, Cuzin N, Traore AS, Garcia JL (1992) Anaerobic degradation of 1,2-propanediol by a new *Desulfovibrio* strain and *D. alcoholovorans*. Arch Microbiol 158:218–225

Pillard DA (1995) Comparative toxicity of formulated glycol deicers and pure ethylene and propylene glycol to *Ceriodaphnia dubia* and *Pimephales promelas*. Environ Toxicol Chem 14:311–315

Pillard DA, DuFresne DL (1999) Toxicity of formulated glycol deicers and ethylene and propylene glycol to *Lactuca sativa*, *Lolium perenne*, *Selenastrum capricornutum*, and *Lemna minor*. Arch Environ Contam Toxicol 37:29–35

Price K, Waggy GT, Conway RA (1974) Brine shrimp bioassay and seawater BOD of petrochemicals. J Water Pollut Control Fed 46:63–77

Puck WS, Tamplin WS (1952) Physical properties of propylene glycol. In: Curme GO, Johnston F (eds) Glycols. Reinhold Publishing Company, New York, NY

Reynolds T (1977) Comparative effects of aliphatic compounds on inhibition of lettuce fruit germination. Ann Bot 41:637–648

Roberts DW, Costello JF (2003) Mechanisms of action for general and polar narcosis: a difference in dimension. QSAR Comb Sci 22:226–233

Sabljic A, Gusten H, Verhaar H, Hermens J (2005) QSAR modeling of soil sorption. Improvements and systematics of log Koc vs. log Kow correlations. Chemosphere 31:4489–4514

SETAC (1993) Guidance document on sediment toxicity tests and bioassays for freshwater and marine environments. In: Hill I, Mathiessen P, Heimbach F (eds) Workshop on sediment toxicity assessment, Renesse, Netherlands, 8–10 Nov 1993. Society of Environmental Toxicology and Chemistry, Brussels

Sezgin N, Tomuk GU (2013) Anaerobic treatability of wastewater contaminated with propylene glycol. Bull Environ Contam Toxicol 91:320–323

Sills RD, Blakeslee PA (1992) The environmental impact of deicers in airport stormwater runoff. In: D'Itri FM (ed) Chemical deicers and the environment. Lewis, Boca Raton, FL, pp 323–340

Sullivan CJ (1993) Propanediols. In: Ullmann's encyclopedia of industrial chemistry, vol A-22, 5th edn. VCH, Deerfield Beach, FL

Tarkpea M, Hansson M, Samuelsson B (1986) Comparison of the microtox test with the 96-hour LC_{50} test for the harpacticoid *Nitocra spinies*. Ecotoxicol Environ Saf 11:127–143

The Dow Chemical Company (2003) A guide to glycols. The Dow Chemical Company, Midland, MI, Available at: http://www.dow.com/PublishedLiterature

The Dow Chemical Company (2013) Dow UCAR™ aircraft deicing and anti-icing fluids: delivering on-time flight performance safety. The Dow Chemical Company, Midland, MI, Available at: http://www.dow.com/aircraft/index.htm

Tsuji S, Tonogai Y, Ito Y, Kanoh S (1986) The influence of rearing temperatures on the toxicity of various environmental pollutants for killifish (*Oryzias latipes*). J Hyg Chem (Eisei Kagaku) 32:46–53

Turoski VE, Woltman DL, Vincent BF (1983) Determination of organic priority pollutants in the paper industry by GC/MS. Tappi J 66:89–90

USEPA (2000) Preliminary data summary: airport deicing operations (Revised). Office of Water, United States Environmental Protection Agency, Washington, DC, EPA-821-R-00-016

USEPA (2006) Reregistration eligibility decision document—propylene glycol and dipropylene glycol. Office of Pesticide Programs, United States Environmental Protection Agency, Washington, DC, EPA-739-R-06-002

USEPA (2012) Estimation programs interface for windows (EPI Suite), version 4.10. Office of Pollution Prevention and Toxics, United States Environmental Protection Agency, Washington, DC, Available at: http://www.epa.gov/oppt/exposure/pubs/episuite.htm

Verschueren K (2001) Handbook of environmental data on organic chemicals, 4th edn. Wiley, New York, NY

Weast RC, Astle MJ (1985) Handbook of data on organic compounds. CRC Press LLC, Boca Raton, FL

West RJ, Davis JW, Pottenger LH, Banton MI, Graham C (2007) Biodegradability relationships among propylene glycol substances in the organization for economic cooperation and development ready- and seawater biodegradability tests. Environ Toxicol Chem 26:862–871

Zgola-Grzeskowiak A, Grzeskowiak T, Zembrzuska J, Lukaszewski Z (2006) Comparison of biodegradation of poly(ethylene glycol)s and poly(propylene glycol)s. Chemosphere 64:803–809

Zgola-Grzeskowiak A, Grzeskowiak T, Zembrzuska J, Franska M, Franski R, Kozik T, Lukaszewski Z (2007) Biodegradation of poly(propylene glycol)s under the conditions of the OECD screening test. Chemosphere 67:928–933

Zgola-Grzeskowiak A, Grzeskowiak T, Zembrzuska J, Franska M, Franski R, Lukaszewski Z (2008) Bio-oxidation of tripropylene glycol under aerobic conditions. Biodegradation 19:365–373

Index

A

Abiotic degradation of chlorothalonil, hydrolysis, **232**: 93
Activated carbon, heavy metal adsorbent, **232**: 64
Agricultural soil responses, fly-ash amendment, **232**: 45 ff.
Agricultural waste adsorbents for heavy metals, performance parameters (table), **232**: 67–68
Agricultural waste, heavy metal adsorbents, **232**: 65
Agricultural wastes, composition described, **232**: 66
Air chemodynamics, chlorothalonil, **232**: 93
Antioxidant enzymes, plant defense role, **232**: 21
Aquatic degradation pathway, chlorothalonil (diag.), **232**: 94
Aquatic organism effects, chlorothalonil, **232**: 99
Aquatic species toxicity, chlorothalonil (table), **232**: 99
Aquatic species toxicity, dipropylene glycol (table), **232**: 130
Aquatic species toxicity, monopropylene glycol (table), **232**: 127–129
Aquatic species toxicity, tri- & tetra-propylene glycols (tables), **232**: 131
Aquatic species, chlorothalonil bioaccumulation, **232**: 99
Atmospheric half-life, propylene glycol substances (table), **232**: 115
Atmospheric transport, propylene glycol substances, **232**: 119

B

Bioaccumulation of chlorothalonil, aquatic species, **232**: 99
Bioaccumulation, propylene glycol substances, **232**: 125
Biodegradation summary, propylene glycol substances (table), **232**: 122–3
Biodegradation, propylene glycol substances, **232**: 120
Biological responses of soil, to fly ash amendment (table), **232**: 50
Biological soil effects, fly ash amendment, **232**: 49
Biotic breakdown, chlorothalonil, **232**: 96
Bird toxicity, chlorothalonil, **232**: 100

C

Carbohydrate damage in plants, heavy metal exposure, **232**: 13
Cell signaling interference, heavy metals, **232**: 14
Chemical changes in soil, from fly ash amendment, **232**: 50
Chemical characteristics, fly ash, **232**: 48
Chemical composition, oil palm biomass (table), **232**: 69
Chemical modification, oil palm biomass, **232**: 72
Chemical treatments, to enhance oil palm adsorption of heavy metals, **232**: 75
Chemical treatments, to enhance oil palm-based adsorbents (table), **232**: 74
Chemistry, chlorothalonil, **232**: 90
Chemodynamics, chlorothalonil, **232**: 91

Chlorothalonil bioaccumulation, in aquatic species, **232**: 99
Chlorothalonil breakdown pathway, Fenton-reagent induction (diag.), **232**: 95
Chlorothalonil toxicity, aquatic species, **232**: 99
Chlorothalonil toxicity, to aquatic species (table), **232**: 99
Chlorothalonil toxicity, to birds, **232**: 100
Chlorothalonil toxicity, to mammals, **232**: 100
Chlorothalonil toxicity, to plants & fungi, **232**: 101
Chlorothalonil, aquatic degradation pathway (diag.), **232**: 94
Chlorothalonil, aquatic organism effects, **232**: 99
Chlorothalonil, biotic breakdown, **232**: 96
Chlorothalonil, chemical structure (illus.), **232**: 90
Chlorothalonil, chemistry, **232**: 90
Chlorothalonil, chemodynamics in air, **232**: 93
Chlorothalonil, chemodynamics, **232**: 91
Chlorothalonil, environmental degradation, **232**: 93
Chlorothalonil, environmental fate & toxicity, **232**: 89 ff.
Chlorothalonil, microbial degradation pathways (diag.), **232**: 97
Chlorothalonil, microbial degradation products (table), **232**: 98
Chlorothalonil, photolysis, **232**: 94, 96
Chlorothalonil, physiochemical properties (table), **232**: 91
Chlorothalonil, soil adsorption & degradation, **232**: 91
Chlorothalonil, soil chemodynamics, **232**: 91
Chlorothalonil, soil degradation pathway (diag.), **232**: 98
Chlorothalonil, soil leaching potential, **232**: 92
Chlorothalonil, soil runoff, **232**: 91, 92
Chlorothalonil, toxic mode of action, **232**: 97
Chlorothalonil, toxicology, **232**: 97
Chlorothalonil, water chemodynamics, **232**: 92
Chorothalonil, nature, uses & history described, **232**: 89
Coal use, in India, **232**: 46
Contamination, heavy metal sources (table), **232**: 63

D

Degradation half-lives, propylene glycol substances (table), **232**: 115
Dipropylene glycol toxicity, aquatic & terrestrial species (table), **232**: 130

Dipropylene glycol, ecotoxicity, **232**: 126
DNA damage in plants, heavy metals, **232**: 11

E

Ecotoxicity, mono- & di-propylene glycols, **232**: 126
Ecotoxicity, tri- & tetra-propylene glycols, **232**: 130
Endocrine disruption potential, propylene glycol substances, **232**: 132
Environmental compartments, relevant to propylene glycol substances, **232**: 114
Environmental contamination, heavy metal sources (table), **232**: 63
Environmental degradation, chlorothalonil, **232**: 93
Environmental distribution, fate & effects, propylene glycol substances, **232**: 107 ff.
Environmental distribution, propylene glycol substances (table), **232**: 116
Environmental distribution, propylene glycol substances, **232**: 114
Environmental fate, chlorothalonil, **232**: 89 ff.
Environmental fate, propylene glycol substances, **232**: 119
Environmental monitoring data, propylene glycol substances, **232**: 117
Environmental residence times, propylene glycol substances (table), **232**: 116

F

Fenton reaction in plants, ROS production (diag.), **232**: 7
Fenton-reagent-induced breakdown, of chlorothalonil (diag.), **232**: 95
Fly ash amendment, biological effects in soil, **232**: 49
Fly ash amendment, effects on soil chemistry, **232**: 50
Fly ash amendment, soil enzyme implications, **232**: 53
Fly ash amendment, soil responses (table), **232**: 50
Fly ash management, soil biochemical cycle, **232**: 52
Fly ash management, soil microbial dynamics, **232**: 53
Fly ash physico-chemical effects, in soil, **232**: 49
Fly ash, annual production & utilization, **232**: 46
Fly ash, composition & chemical characteristics, **232**: 48
Fly ash, described, **232**: 46, 47

Index

Fly ash, essential elements for plant growth, **232**: 48
Fly ash, physico-chemical properties, **232**: 47
Fly ash, production and utilization (diag.), **232**: 47
Fly ash, properties vs. soil (table), **232**: 49
Fly ash, radionuclide content, **232**: 49
Fly-ash amendment, agricultural soil responses, **232**: 45 ff.
Fly-ash amendment, soil health responses, **232**: 54
Fungi, chlorothalonil toxicity, **232**: 101

G

Genomics, microbial soil dynamic implications, **232**: 54
Glutathionylation in plants, defense against heavy-metal toxicity, **232**: 18

H

Haber-Weiss pathways in plants, ROS production (diag.), **232**: 7
Heavy- metal genotoxic effects, to plants, **232**: 11
Heavy metal adsorbent performance, thermal modification of oil palm biomass (table), **232**: 77–78
Heavy metal adsorbent performance, unmodified oil palm biomass (table), **232**: 71
Heavy metal adsorbent, oil palm biomass, **232**: 61 ff., 69
Heavy metal adsorbents, activated carbon & alumina, **232**: 64
Heavy metal adsorbents, agricultural waste, **232**: 65
Heavy metal adsorbents, commercial options described, **232**: 64
Heavy metal adsorbents, examples described, **232**: 66
Heavy metal adsorbents, zeolite & silica gel, **232**: 65
Heavy metal effects in plants, cell signaling, **232**: 14
Heavy metal effects on plants, protein damage, **232**: 12
Heavy metal effects, on plants, **232**: 3
Heavy metal pollutants, definition & description, **232**: 2
Heavy metal treatment, methods described, **232**: 62
Heavy metals & metalloids, annual production changes (diag.), **232**: 2

Heavy metals contamination, sources & toxic effects (table), **232**: 63
Heavy metals, nature & sources described, **232**: 62
Heavy metals, ROS induction, **232**: 3
Heavy-metal adsorbent performance parameters, agricultural waste (table), **232**: 67–68
Heavy-metal adsorbent performance, chemical modification of oil palm biomass (table), **232**: 74
Heavy-metal effects, on plant lipid peroxidation, **232**: 9
Heavy-metal induction, ROS in plants, **232**: 5
Heavy-metal removal method, adsorption, **232**: 64
Heavy-metal toxicity in plants, nitrogen metabolism role, **232**: 20
Heavy-metal-induced lipid peroxidation, in plants (diag.), **232**: 9
Heavy-metal-induced oxidative stress, scheme in plants (illus.), **232**: 23
Heavy-metal-induced reactive oxygen species, phytotoxicity, **232**: 1ff.
Heavy-metal-induced ROS, in plant species (table), **232**: 6
Heavy-metal-induced ROS, plant defense enzymes (table), **232**: 22
Heavy-metal-induced ROS, plant tolerance mechanisms, **232**: 14
Heavy-metal-induced stress, plants, **232**: 3
Heavy-metal-induced toxicity, on plant macromolecules, **232**: 8
Hydrolysis, chlorothalonil, **232**: 93
Hydrolysis, propylene glycol substances, **232**: 124

I

India, coal use, **232**: 46

L

Lipid peroxidation in plants, induced by heavy metals (diag.), **232**: 9
Lipid peroxidation, heavy-metal effect in plants, **232**: 9

M

Mammalian toxicity, chlorothalonil, **232**: 100
Microbial degradation pathways, chlorothalonil (diag.), **232**: 97

Microbial degradation products, chlorothalonil (table), **232**: 98
Microbial soil dynamic implications, genomics, **232**: 54
Mode of action, chlorothalonil toxicity, **232**: 97
Monopropylene glycol toxicity, aquatic & terrestrial species (table), **232**: 127–129
Monopropylene glycol, ecotoxicity, **232**: 126

N

Nitrogen metabolism in plants, role in heavy-metal toxicity, **232**: 20

O

Oil palm adsorbents for heavy metals, chemical modification effects (table), **232**: 74
Oil palm adsorbents for heavy metals, thermal modification effects (table), **232**: 77–78
Oil palm adsorbents for heavy metals, unmodified biomass performance parameters (table), **232**: 71
Oil palm adsorption performance, modified biomass, **232**: 72
Oil palm adsorption performance, unmodified biomass, **232**: 70
Oil palm biomass adsorption effects, thermal treatments (table), **232**: 77–78
Oil palm biomass as heavy metal adsorbent, future research needs, **232**: 79
Oil palm biomass performance, from chemical modification, **232**: 72
Oil palm biomass performance, from thermal modification, **232**: 73
Oil palm biomass, chemical composition (table), **232**: 69
Oil palm biomass, heavy metal adsorbent, **232**: 61 ff.
Oil palm biomass, heavy metal adsorbent, **232**: 69
Oil palm, source, history & production, **232**: 69
Oxidative stress scheme in plants, heavy-metal-induced (illus.), **232**: 23

P

Photolysis, chlorothalonil, **232**: 94, 96
Physico-chemical changes in plants, heavy-metal-induced ROS, **232**: 1ff.
Physico-chemical effects of fly ash, in soil, **232**: 49

Physico-chemical properties, fly ash (table), **232**: 49
Physico-chemical properties, fly ash, **232**: 47
Physico-chemical properties, propylene glycol substances (table), **232**: 110
Physico-chemical properties, propylene glycol substances, **232**: 109
Physiochemical properties, chlorothalonil (table), **232**: 91
Phytotoxicity, heavy-metal-ROS, **232**: 1ff.
Plant damage from heavy metals, carbohydrates, **232**: 13
Plant damage from heavy metals, proteins, **232**: 12
Plant defense enzymes, heavy-metal-induced ROS (table), **232**: 22
Plant defense mechanisms, against ROS injury, **232**: 3
Plant defense, role of antioxidant enzymes, **232**: 21
Plant effects, of heavy metals, **232**: 3
Plant effects, of lipid peroxidation, **232**: 10
Plant effects, of ROS, **232**: 3
Plant macromolecules, ROS effects, **232**: 8
Plant metabolic roles, ROS, **232**: 7
Plant metabolism, ROS production, **232**: 4
Plant nitrogen metabolism, role in heavy-metal toxicity, **232**: 20
Plant production of ROS, by heavy metals, **232**: 5
Plant production of ROS, natural generation, **232**: 4
Plant species, heavy-metal-induced ROS (table), **232**: 6
Plant stress induction scheme, by heavy metals (illus.), **232**: 23
Plant tolerance mechanisms, to heavy metals, **232**: 14
Plant toxicity, chlorothalonil, **232**: 101
Plant toxicity, primary defense against heavy metals, **232**: 15
Plant toxicity, secondary defenses against heavy metals, **232**: 16
Propylene glycol substance half-lives, environmental matrices, **232**: 115
Propylene glycol substances, atmospheric transport, **232**: 119
Propylene glycol substances, bioaccumulation, **232**: 125
Propylene glycol substances, biodegradation summary (table), **232**: 122–123
Propylene glycol substances, biodegradation, **232**: 120

Propylene glycol substances, degradation half-lives (table), **232**: 115
Propylene glycol substances, density/specific gravity, **232**: 110
Propylene glycol substances, endocrine disruption potential, **232**: 132
Propylene glycol substances, environmental distribution, fate & effects, **232**: 107 ff.
Propylene glycol substances, environmental fate, **232**: 119
Propylene glycol substances, environmental monitoring data, **232**: 117
Propylene glycol substances, Henry's Law constant, **232**: 112
Propylene glycol substances, hydrolysis, **232**: 124
Propylene glycol substances, identity (table), **232**: 108
Propylene glycol substances, melting/freezing & boiling points, **232**: 111
Propylene glycol substances, modeled environmental distribution & residence times (table), **232**: 116
Propylene glycol substances, nature, uses, production described, **232**: 108
Propylene glycol substances, octanol-water partition coefficient (Log P_{ow}), **232**: 113
Propylene glycol substances, organic carbon-normalized adsorption coefficient (Log K_{oc}), **232**: 113
Propylene glycol substances, physico-chemical properties (table), **232**: 110
Propylene glycol substances, physico-chemical properties, **232**: 109
Propylene glycol substances, relevant environmental compartments, **232**: 114
Propylene glycol substances, vapor pressure, **232**: 111
Propylene glycol substances, water solubility, **232**: 112
Protein damage in plants, heavy metal exposure, **232**: 12

R

Radionuclide content, fly ash, **232**: 49
Reactive oxygen species (ROS), phytotoxic effects, **232**: 1ff.
ROS (reactive oxygen species) induction, by heavy metals, **232**: 3
ROS damage in plants, to proteins and carbohydrates, **232**: 12, 13
ROS effects, on plant macromolecules, **232**: 8
ROS genotoxicity in plants, from heavy metals, **232**: 11
ROS in plant species, heavy-metal-induced (table), **232**: 6
ROS in plants, antioxidant enzyme defenses, **232**: 21
ROS injury mechanisms, plant defenses, **232**: 3
ROS production in plants, by heavy metals, **232**: 5
ROS production in plants, Haber-Weiss & Fenton pathways (diag.), **232**: 7
ROS production, in plant metabolism, **232**: 4
ROS roles, in plant metabolism, **232**: 7
ROS, defined, **232**: 4
ROS, effects on plants, **232**: 3
ROS, natural plant production, **232**: 4

S

Sediment degradation half-lives, propylene glycol substances (table), **232**: 115
Silica gel, heavy metal adsorbent, **232**: 65
Soil adsorption & degradation, chlorothalonil, **232**: 91
Soil amendment by fly ash, biological responses (table), **232**: 50
Soil biochemical cycle, fly ash management, **232**: 52
Soil chemistry changes, from fly ash amendment, **232**: 50
Soil chemodynamics, chlorothalonil, **232**: 91
Soil degradation half-lives, propylene glycol substances (table), **232**: 115
Soil degradation pathway, chlorothalonil (diag.), **232**: 98
Soil enzyme implications, of fly ash amendment, **232**: 53
Soil health responses, to fly-ash amendment, **232**: 54
Soil leaching potential, chlorothalonil, **232**: 92
Soil microbial dynamics, fly ash management, **232**: 53
Soil responses, fly-ash amendment, **232**: 45 ff.
Soil runoff, chlorothalonil, **232**: 91, 92

T

Terrestrial species toxicity, monopropylene glycol (table), **232**: 127–129
Terrestrial species toxicity, dipropylene glycol (table), **232**: 130

Terrestrial species toxicity, tri- & tetra-propylene glycols (tables), **232**: 131
Thermal modification, oil palm biomass, **232**: 73
Toxic effects, heavy metals (table), **232**: 63
Toxic mode of action, chlorothalonil, **232**: 97
Toxicity of heavy-metal-induced ROS, on plant macromolecules, **232**: 8
Toxicity to aquatic species, chlorothalonil, **232**: 99
Toxicity to mammals & birds, chlorothalonil, **232**: 100
Toxicity to plants & fungi, chlorothalonil, **232**: 101
Toxicity, chlorothalonil to aquatic species (table), **232**: 99
Toxicology, chlorothalonil, **232**: 89 ff., 97
Tri- & tetra-propylene glycol toxicity, aquatic & terrestrial species (tables), **232**: 131
Tri- & tetra-propylene glycols, ecotoxicity, **232**: 130

W

Water degradation half-lives, propylene glycol substances (table), **232**: 115

Z

Zeolite, heavy metal adsorbent, **232**: 65